research and education in mathematics

Edited by

Karl H. Hofmann
Dept. of Mathematics
Tulane University
New Orleans, LA 70118
U.S.A.

Rudolf Wille
Fachbereich Mathematik
Techn. Hochschule Darmstadt
Schloßgartenstr. 7
D - 6100 Darmstadt
Fed. Rep. Germany

Titles in this Series

Volume 1 — R.T.Rockafellar: The Theory of Subgradients and its Applications to Problems of Optimization. Convex and Nonconvex Functions

Volume 2 — J.Dauns: A Concrete Approach to Division Rings

Volume 3 — L.Butz: Connectivity in Multi-factor Designs

Instructions for Authors are given on the backside cover.

R & E 3
research and education in mathematics
edited by Karl H. Hofmann and Rudolf Wille

Lothar Butz

Connectivity in Multi-factor Designs

A Combinatorial Approach

Heldermann Verlag Berlin

Lothar Butz
Institut für Ökonometrie
und Operations Research
Universität Bonn
Nassestraße 2
D - 5300 Bonn 1
Fed. Rep. Germany

AMS Subject Classification Scheme 1979/80: 62Kxx; 05-02, 05B05, 05C20, 05C40, 62-02, 62P20, 68E10, 90-02, 90A19, 90A20, 90B99, 94C15, 94C30.

Deutsche Bibliothek Cataloguing in Publication Data

Butz, Lothar:

Connectivity in multi-factor designs : a combinatorial approach / Lothar Butz. - Berlin : Heldermann, 1982.
(Research and education in mathematics ; Vol. 3)
ISBN 3-88538-203-2
NE: GT

ISBN 3-88538-203-2

To my parents

ACKNOWLEDGEMENTS

I am very much indebted to Bernhard Korte who has always supported and fostered my work. Also I am greatly obliged to Rabe von Randow for improving the translation and last but not least to Harry Baumann who did a very good typing job.

L.B.

CONTENTS

1. Outline 1

2. Preliminaries, problem formulation and survey of known results 7

2.1 The linear model 7

2.2 Graphs 22

2.3 State of the art 25

3. Characterizations of connected designs by means of digraphs with labeled cycles 39

3.1 Prerequisites 39

3.2 Row-column designs 51

3.3 Designs for two-way elimination of heterogeneity 59

3.4 Four-factor designs 113

3.5 Multi-factor designs 135

4. Invariance properties, reduction methods, algorithms 155

List of symbols 183

References 185

Index 189

1. OUTLINE

During the last decades the theory of experimental design has given rise to an extensive set of methods which can be applied successfully for the solution of many problems arising in economics, technology, medicine and agricultural science. In the last few years there have also been attempts to characterize certain concepts of experimental design by combinatorial means, in particular by graph theoretical methods. Most of these investigations were restricted to rather special cases as the methods used could not be adapted to more general situations. In the present monograph experimental designs are treated under the aspect of connectivity (cf. section 2.1 for the precise definition). Starting with special cases a stepwise increase in generality will introduce the reader gradually to the formally more involved techniques applicable to problems of arbitrary generality. The framework best suited for the formulation of these techniques is that of graph theory.

Simply stated, the experimental design problem considered here is that of analysing the effects of certain *treatments* on a set of given experimental units within the underlying statistical model, by designing a series of experiments. The property of *connectivity* proves to be an essential concept in this context: We shall see that connected resp. completely connected designs offer the best possible estimability conditions for the parameters of the underlying statistical model. In addition to the direct applicability of the combinatorial criteria presented, they also have interesting theoretical consequences.

The notion of connectivity goes back to a concept developed by Bose [1947] for block designs (two-factor designs), which yields a characterization of the estimability of a certain class of linear functions (so-called treatment contrasts) in the parameters of the linear model under consideration. The concept of *blocks* in experimental designs was introduced by Yates [1936] in order to take account of the effects of perturbations in the analysis of designs caused by heterogeneous experimental conditions. For this purpose one partitions the experimental units of the design into blocks in such a way that units of the same block are always affected by the same perturbation. If each experimental unit is affected by several perturbation factors (*block factors*), their effects can be eliminated by multiple block partitioning.

We now give a brief outline of the results presented. In the first section of the next chapter the connectivity property of a design with $n \geqq 2$ factors is defined within the framework of the underlying statistical model. We derive fundamental characterizations of estimable linear functions in the parameters of the model which later on turn out to be very useful for proving results on connectivity. The second section consists of a short survey of the graph theoretical concepts needed. Section 2.3 reviews the state of the art in the field of connectivity characterizations. The prevalent approach in the literature to block designs and to designs for two-way elimination of heterogeneity characterizes connectivity via a rank condition on the so-called C-matrix (cf. e.g. Chakrabarti [1962]). It is comparatively easy to calculate the associated C-matrix in the case of a block design or a *row-column design*. For general designs with three factors the calculation of the C-matrix involves a generalized inverse which is not simple to compute (cf. Raghavarao/ Federer [1975]). Only for a special class of designs with more than three factors has it been shown how the C-matrix can be determined (Cheng [1978]). Two elegant combinatorial characterizations of connectivity were published a few years ago, namely for block designs (by means of the connectedness of an associated bipartite graph, cf. Gaffke [1978]) and for row-column designs with only two rows (via the connectedness of an associated digraph and an additional condition involving a cycle in this digraph, cf. Wynn [1977]). These two results, which are presented with new proofs in section 2.3, motivated the author to work on this subject.

The main results are contained in the third chapter. The first section provides several lemmas which are needed for the proofs in the following sections. Also a first necessary and sufficient condition for the connectivity of arbitrary multi-factor designs in the form of a rank criterion is presented (lemma (3.1.2)). Section 3.2 contains a combinatorial characterization of connectivity for row-column designs with arbitrarily many rows and columns. For this purpose a uniquely determined digraph with integer labels on the arcs is associated with a given row-column design. A necessary and sufficient connectivity criterion for the row-column design can then be formulated using the connectedness of this digraph and a condition on the cycles involving the arc labels. In section 3.3 general three-factor designs are dealt with in detail. It is shown how the definition of the digraphs introduced for row-column designs can be modified appropriately. Here the condition on the cycles also involves the number of connected components of a further graph given by the design in question. Altogether six equivalent connectivity characteriaztions for general three-factor designs are given which are all based on different digraphs with labeled arcs. This makes it possible to minimize the effort needed to check a connectivity criterion in the sense that one has the choice of a suitable digraph depending on the dimensions of the three-factor design to be studied. In many cases it is immediately clear from the associated digraph whether the underlying design is connected or not: if for one of the three factors of the experimental design only a small number of different levels has to be considered, then the cycle condition can be checked very easily, in the case of only two or three levels even by mere inspection of the digraph. But in any case the effort needed compares very favourably with the calculation of the corresponding C-matrix and the computation of its rank. For four of the six digraphs introduced for the characterization of connectivity, the graph theoretical connectedness happens to be a necessary condition, and this can be checked for designs of any size without difficulty.

Further results of this section relate the connectivity of a three-factor design to the connectivity of its induced two-factor designs (i. e. the designs which are obtained by simply ignoring any one of the three-factors):
if in a connected three-factor design one of the two blocking factors is disregarded, then a two-factor design results which is also connected. If on the other hand the treatment factor is ignored, then the resulting two-factor design is connected if and only if the three-factor design in question is *completely* connected (i. e. besides estimability of all treatment contrasts the estimability of all blocking factor contrasts is also guaranteed). For two-factor designs the properties of being *connected* and *completely connected* agree, and for general three-factor designs one can modify the connectivity criteria accordingly so that characterizations for complete connectivity result. Finally, the proof techniques employed in this context made it possible to show the equivalence of several distinct estimability criteria for the complete connectivity of a design.

The next two sections deal with generalizations of the results obtained for three-factor designs to the case of four- resp. multi-factor designs; in order to facilitate the presentation, the investigation of general n-factor designs has deliberately been postponed to the end of the chapter. Whereas three-factor designs were discussed in full detail, section 3.4 is restricted to aspects which are of particular interest for four-factor designs. Again connectivity characterizations by means of digraphs with labeled arcs are given, the cycle condition now making use of the digraphs introduced for three-factor designs in the preceding section.

Finally in section 3.5 the connectivity property of designs with arbitrarily many factors is characterized by associated digraphs of corresponding generality. Clearly the definitions required will be correspondingly by more complicated than before, although the combinatorial approach again proves to be fruitful. The third chapter concludes with a number of results on the connectivity properties of designs induced by a connected multi-factor design, as well as on equivalent estimability characterizations of complete connectivity.

The last chapter deals with the algorithmic treatment of the criteria developed in the third chapter. Invariance properties and reduction methods are discussed which allow the stepwise simplification of digraphs. Two simple reduction rules for arc labeled digraphs are deduced and then used to demonstrate the efficacy of the combinatorial approach. Even very large designs can be analysed by elementary methods by taking advantage of certain structural properties. Furthermore an algorithm for the stepwise evaluation of the crucial cycle condition is presented and its time complexity estimated. In this algorithm the analysis of the corresponding digraph is carried out implicitly by successively evaluating the single experimental units.

It is evidently of considerable importance that the connectivity property can now be tested even in the case of those multi-factor designs for which no, or at least no practicable, characterization was known. Furthermore the combinatorial view of the concept of connectivity proves to be very useful in applications. Thus the digraphs introduced for the characterizations of connectivity offer new insight into the structure of the matrices representing connected designs. This facilitates the construction of examples of connected designs with partially fixed data. One is for example often faced with the problem of extending a given (non-connected) design to a connected one by adding further suitable experiments or to reduce the number of experiments of a connected design without forfeiting the connectivity property - both problems can easily be solved by means of the associated digraphs.

2. PRELIMINARIES, PROBLEM FORMULATION AND SURVEY OF KNOWN RESULTS

2.1 The linear model

As a framework for the experimental designs to be dealt with, consider a *linear statistical model* (cf. Krafft [1978], Searle [1971]) of the form

$$x = B\pi + \varepsilon \ . \tag{2.1.1}$$

x is a p-dimensional random vector whose components denote the observable outcomes of the experiments comprising the design to be described. π is a q-dimensional vector of unknown parameters about which one whishes to obtain information by means of the experiments. ε is a p-dimensional random vector whose components measure the noise and the observational errors in the experiments. B is a $p \times q$ 0,1-matrix by which an *experimental design* is defined. The purpose of such an experimental design is to study the effects of a given set of n different *factors* $F_1, \dots, F_n$ influencing the experimental setup, where each F_i has m_i different *levels* ($1 \leq i \leq n$, $m_i \geq 2$). The experimental design consists of p experiments given by the p rows of B. Exactly one level of each factor F_i occurs in each experiment. The parameter vector π contains exactly one component for each level of every factor. Accordingly π is subdivided into vectors

$$\pi^i = (\pi^i_1, \dots, \pi^i_{m_i})^T$$

corresponding to the factors F_i where $(\ )^T$ denotes transposition of vectors and matrices, i. e.

$$\pi = ((\pi^1)^T, \dots, (\pi^n)^T)^T .$$

Writing B in the form

$$B = (B^1, \dots, B^n)$$

where each $p \times m_i$ submatrix B^i is the appropriate block of columns for factor F_i, we have $\sum_{i=1}^{n} m_i = q$ and $B\pi = \sum_{i=1}^{n} B^i \pi^i$. The columns of the matrix B^i correspond to the levels of factor F_i, $1 \leq i \leq n$. Every row of each matrix B^i is an m_i-dimensional unit vector, the non-zero elements of a given row of B determine the combination of factor levels involved in the corresponding experiment.

Consequently, the p experiments of the design given by B are defined as follows: The k-th experiment is set up by combining the level j_1 of factor F_1, level j_2 of factor F_2 and so on through level j_n of factor F_n, if and only if the k-th row vector of B^i equals the j_i-th unit vector of $\mathbb{R}^{m_i}$ (m_i-dimensional real space) for $i = 1, \dots, n$ $(1 \leq k \leq p)$.

It is assumed that ε (vector of errors) has zero expectation. Hence a single equation from the model (2.1.1) implies

(2.1.2) $$\mathcal{E}(x_{j_1 \dots j_n d}) = \sum_{i=1}^{n} \pi^i_{j_i}$$

where $\mathcal{E}$ denotes the expectation operator and the indices in (2.1.2) are taken to indicate that the experiment in question is the d-th experiment in which for $i = 1, \dots, n$ level j_i of factor F_i is applied. Obviously

$1 \leq j_i \leq m_i$ for all $i \in \{1,\dots,n\}$. Taking $d_{j_1 \dots j_n}$ to be the number of experiments in the design with the same combination $j_1,\dots,j_n$ of factor levels, the index d in (2.1.2) must satisfy $1 \leq d \leq d_{j_1 \dots j_n}$. From (2.1.2) it is also clear that a purely additive model without interactions between the effects of different factors is considered.

In what follows the rows of B are always considered to be arranged according to the lexicographical order of the index tuples $(j_1,\dots,j_n,d)$, unless specified otherwise. This assumption guarantees that B is uniquely determined by means of the $m_1 \times \dots \times m_n$ "matrix"

$$D = ((d_{j_1 \dots j_n})) .$$

Generalizing the usual notion of (real) two-dimensional $m_1 \times m_2$ matrices, which can be conceived as mappings of the cartesian product $\{1,\dots,m_1\} \times \{1,\dots m_2\}$ into the real numbers, an $m_1 \times \dots \times m_n$ matrix M is taken to be a mapping

$$M : \mathop{X}_{i=1}^{n} \{1,\dots,m_i\} \longrightarrow \mathbb{R} .$$

<u>2.1.3 Definition</u>

Let $n,m_1,\dots,m_n \geq 2$. The experimental design given by a non-negative integer $m_1 \times \dots \times m_n$ matrix D is called an *n-factor design* (also denoted by D to simplify matters).

2.1.4 Remark

Without loss of generality we consider only designs D for which each level of every factor occurs in at least one of the experiments, i. e.

$$\prod_{i=1}^{n} \prod_{j_i=1}^{m_i} \left(\sum_{\substack{k=1\\k\neq i}}^{n} \sum_{j_k=1}^{m_k} d_{j_1 \dots j_n} \right) > 0 .$$

(This can easily be achieved by eliminating "superfluous" factor levels if necessary.)

In many applications one is only interested in the effects of one single factor. In this case the other factors are only taken into consideration in order to account for distortions caused by heterogeneous conditions of the experiments. This distinguished factor is then called the *treatment factor* and its levels are the *treatments* as they are under the direct control of the experimenter. The remaining factors are called *blocking factors* and their levels *blocking levels* . For our purposes F_n will usually play the role of the treatment factor.

A two-factor design is simply called a *block design* , this notion was introduced by Yates [1936]. If two blocking factors are considered, the resulting three-factor design is a so-called *design for two-way elimination of heterogeneity* . Special cases which have been studied frequently are designs in which there is exactly one experiment for each combination of blocking levels, i. e.

$$\sum_{k=1}^{m_3} d_{ijk} = 1 \quad \text{for all} \quad i \in \{1,\dots,m_1\},\ j \in \{1,\dots,m_2\}.$$

Three-factor designs D of this kind are usually called *row-column designs*, as D can then be uniquely represented by an $m_1 \times m_2$ matrix $Y = ((y_{ij}))$, $y_{ij} \in \{1,\dots,m_3\}$, where

$$y_{ij} = k \iff d_{ijk} = 1 .$$

Here the *row* indices i $(1 \leq i \leq m_1)$ correspond to the levels of the first blocking factor and the *column* indices j $(1 \leq j \leq m_2)$ to the levels of the second blocking factor.

2.1.5 Example of a block design

Consider a project for studying the effect of different alloys on the life span of special tools (e. g. drilling heads). These tools are tested on different tasks (e. g. different sorts of rock occurring in oil well drilling) whose influence on the life span of the tools is also not known. It is assumed that interactions between the choice of material and task are negligible and that the expected life span can be calculated as the sum of two parameters, namely one for the material used and one for the task to be performed. Here however the effect of the material used is crucial whereas the task is of minor importance. This problem can be analysed by means of a block design: the choice of material is the treatment factor and the tasks give rise to the blocking factor.

2.1.6 Example of a row-column design

Examples of row-column designs occur in a natural way in agricultural science: a rectangular experimental field is subdivided like a chess board into lots of equal size, and suppose that blocking levels are assigned to the rows and columns of this arrangement such that the rows correspond to one blocking factor and the columns to the other. Hence the experimental field is an array of the same size $m_1 \times m_2$ as the matrix Y which defines the row-column design. As a practical application consider an experimental design for testing m_3 new breeds of potatoes (treatments) against m_1 sorts of fertilizer and m_2 different amounts of irrigation (to account for climatic effects). Breed k will be planted in the lot with row index i and column index j if and only if $y_{ij} = k$.

2.1.7 Example of a multi-factor design

The Brazilian government pursues a program of developing new kinds of automobile fuel. Its major purpose is to become largely independent of oil imports, therefore alcohol made of sugar cane is the main ingredient. Several fuel mixtures have to be tested for their productiveness. Tests cannot be carried out in huge numbers because of the problem of supply. Heterogeneous experimental conditions have to be taken into account, however these distorting effects (blocking factors) can be eliminated to a large extent by incorporating them in the statistical model. The fuel mixtures to be tested constitute the treatment factor, and one could consider the following blocking factors: types of engines, modes of operation (long distances, short distances), average speed, condition of roads, driving behaviour.

A great number of publications dealing with block designs have appeared, see for instance the articles by Bose [1950/51], Tocher [1952], Kiefer [1959], the books by Raghavarao [1971], Krafft [1978] and their numerous references.

Several publications have been devoted to row-column designs, e. g. Shah/Katri [1973], Federer/Nair/Raghavarao [1975], Pearce [1975], Russell [1976], Jones [1979], cf. also Krafft [1978]. General designs for two-way elimination of heterogeneity have been studied by Shrikhande [1951], Chakrabarti [1962], Kurotschka/Dwyer [1974], Raghavarao/Federer [1975] and Singh/Dey [1978] among others. Only a few articles deal with multi-factor designs, e. g. Srivastava/Anderson [1970], Ecclestone/Russell [1975], Cheng [1978].

Experimental designs are studied in order to obtain information about the vector π of unknown parameters in (2.1.1) . Direct computation or estimation of π or only of the subvector of treatment parameters π^n is not possible, as $x = B\pi + \varepsilon$ cannot have a unique solution in π :

$$\sum_{i=1}^{n} B^i \pi^i = \sum_{i=2}^{n} B^i(\pi^i + \lambda_i \cdot 1_{m_i}) + B^1(\pi^1 - (\sum_{i=2}^{n} \lambda_i) \cdot 1_{m_1})$$

for all $\lambda_1, \dots, \lambda_n \in \mathbb{R}$ ($1_m \in \mathbb{R}^m$ is the m-vector of all ones).

Therefore one examines linear functions of the parameters (cf. Krafft [1978]).

2.1.8 Definition

A linear function $\varphi : \pi \longmapsto f^T\pi$ ($f \in \mathbb{R}^q$) in the parameters of the model $x = B\pi + \varepsilon$ is called *estimable* , if there exists a linear combination of the observations, say $w^T x$ ($w \in \mathbb{R}^p$), such that $f^T\pi = \mathcal{E}(w^T x)$ where $\mathcal{E}$ denotes expectation. $w^T x$ is called a *linear estimation* of φ .

The following lemma yields a simple characterization of estimability which turns out to be of fundamental importance for many results in this context.

2.1.9 Lemma

$\varphi(\pi) = f^T\pi$ is estimable in the model $x = B\pi + \varepsilon$ if and only if φ vanishes for any point π in the kernel of B .

Proof:

$\mathcal{E}(x) = B\pi$ implies

$$\mathcal{E}(w^T x) = w^T B \pi .$$

Hence φ is estimable if and only if there exists a vector $w \in \mathbb{R}^p$ such that

$$f^T\pi = w^T B\pi$$

holds for all π . This is equivalent to

$$\{w \mid B^T w = f\} \neq \emptyset .$$

Now the assertion follows from a well-known variant of Farkas' Lemma (a direct proof makes use of a rank argument).

□

The next lemma contains a necessary condition for the estimability of linear functions in the parameters of an n-factor design.

2.1.10 Lemma

Let

$$\varphi(\pi) = \sum_{i=1}^{n} (f^i)^T \pi^i \quad (f^i \in \mathbb{R}^{m_i})$$

be an estimable function of the parameters in the model (2.1.1), where B is the matrix of an n-factor design D. Then it follows that

$$\sum_{j=1}^{m_i} f_j^i = \sum_{j=1}^{m_{i^*}} f_j^{i^*} \quad \text{for all} \quad i,i^* \in \{1,\dots,n\} .$$

Proof:

Let 0_m denote the m-vector of all zeros. From linear algebra we know that $\exists\, w \in \mathbb{R}^p$ such that $f = B^T w \iff \text{rank } B^T = \text{rank } (B^T,f)$, hence

$$\begin{aligned} \text{rank } B^T B &\leq \text{rank } (B^T B, f) \\ &= \text{rank } (B^T,f) \begin{pmatrix} B & 0_p \\ 0_q^T & 1 \end{pmatrix} \\ &\leq \text{rank } (B^T,f) \\ &= \text{rank } B^T \\ &= \text{rank } B^T B , \end{aligned}$$

i. e.

$$\text{rank}\,(B^TB,f) = \text{rank}\,B^TB\ .$$

Thus

$$\exists\ r \in \mathbb{R}^q \quad \text{such that} \quad B^TB\,r = f\ .$$

The particular structure of the matrix B (each row is composed of n unit vectors) makes it easy to calculate

$$B^TB = \begin{bmatrix} D_1 & D_{12} & \cdots & D_{1i} & \cdots & D_{1n} \\ D_{12}^T & D_2 & \cdots & D_{2i} & \cdots & D_{2n} \\ \vdots & \vdots & & \vdots & & \vdots \\ D_{1i}^T & D_{2i}^T & \cdots & D_i & \cdots & D_{in} \\ \vdots & \vdots & & \vdots & & \vdots \\ D_{1n}^T & D_{2n}^T & \cdots & D_{in}^T & \cdots & D_n \end{bmatrix} ,$$

where the submatrices D_i and D_{ii^*} $(i < i^*)$ are defined as follows:

For $i = 1,\ldots,n$, D_i is the diagonal $m_i \times m_i$ matrix

$$D_i = \text{diag}\,(d_1^i,\ldots d_{m_i}^i)\ ,$$

where the j_i-th diagonal element $d_{j_i}^i$ of D_i is the number of experiments in the design D given by B , in which the j_i-th effect of the i-th factor F_i appears, i. e.

$$d^{i}_{j_i} = \sum_{\substack{\nu=1\\ \nu\neq i}}^{n} \sum_{j_\nu=1}^{m_\nu} d_{j_1\cdots j_i\cdots j_n} \quad , \quad 1 \leqq j_i \leqq m_i \; , \quad 1 \leqq i \leqq n \; .$$

$D_{ii^*} = ((d^{ii^*}_{j_i j_{i^*}}))$ is an $m_i \times m_{i^*}$ matrix where

$$d^{ii^*}_{j_i j_{i^*}} = \sum_{\substack{\nu=1\\ \nu\neq i,i^*}}^{n} \sum_{j_\nu=1}^{m_\nu} d_{j_1\cdots j_i\cdots j_{i^*}\cdots j_n}$$

$$1 \leqq j_i \leqq m_i \; , \quad 1 \leqq j_{i^*} \leqq m_{i^*} \; , \quad 1 \leqq i < i^* \leqq n \; .$$

$d^{ii^*}_{jj^*}$ is the number of experiments in D , in which the j-th level of factor F_i and the j^*-th level of factor F_{i^*} occur together.

Using these concepts one can write the system $B^T B r = f$ as

$$\sum_{k=1}^{i-1} D^T_{ki} r^k + D_i r^i + \sum_{k=i+1}^{n} D_{ik} r^k = f^i \; ,$$

$$1 \leqq i \leqq n \; ,$$

where the vectors r and f have been subdivided into m_i-vectors r^i and f^i , resp., i. e. $r^T = ((r^1)^T, \dots, (r^n)^T)$, $f^T = ((f^1)^T, \dots, (f^n)^T)$. The j_i-th component of the m_i-vector $D_{ik} r^k$ equals

$$\sum_{j=1}^{m_k} d^{ik}_{j_i j} r^k_j \; ,$$

hence the above system of vector equations becomes

$$\sum_{j=1}^{m_i} f_j^i = \sum_{j=1}^{m_i} \sum_{k=1}^{i-1} \sum_{j^*=1}^{m_k} d_{j^*j}^{ki} r_{j^*}^k + \sum_{j=1}^{m_i} d_j^i r_j^i +$$

$$+ \sum_{j=1}^{m_i} \sum_{k=i+1}^{n} \sum_{j^*=1}^{m_k} d_{jj^*}^{ik} r_j^k \quad , \quad 1 \leqq i \leqq n \ .$$

Now

$$\sum_{j=1}^{m_i} d_{jj^*}^{ik} = d_{j^*}^k = \sum_{j=1}^{m_i} d_{j^*j}^{ki}$$

implies

$$\sum_{j=1}^{m_i} f_j^i = \sum_{k=1}^{i-1} \sum_{j=1}^{m_k} d_j^k r_j^k + \sum_{j=1}^{m_i} d_j^i r_j^i + \sum_{k=i+1}^{n} \sum_{j=1}^{m_k} d_j^k r_j^k$$

$$= \sum_{k=1}^{n} \sum_{j=1}^{m_k} d_j^k r_j^k \quad , \qquad 1 \leqq i \leqq n \ .$$

which proves the lemma. □

Using the notation introduced in the proof, the assumption of remark (2.1.4) can be restated as

$$d_j^i > 0 \quad \text{for all} \quad 1 \leqq i \leqq n \ , \quad 1 \leqq j \leqq m_i \ .$$

The following lemma shows that repetitions of the same experiment are irrelevant with respect to the estimability of linear functions in the parameters of a given design.

2.1.11 Lemma

$\varphi(\pi) = f^T\pi$ is estimable with respect to an n-factor design D if and only if $\varphi(\pi)$ is estimable with respect to $D^* = ((d^*_{j_1 \dots j_n}))$ where

$$d^*_{j_1 \dots j_n} = \begin{cases} 1 & \text{if } d_{j_1 \dots j_n} > 0 , \\ 0 & \text{otherwise} . \end{cases}$$

Proof:

Easy consequence of lemma (2.1.9) .

□

As mentioned before, in many situations one is only concerned with the parameters of one single factor F_i :

2.1.12 Definition

$\varphi(\pi) = \sum_{i=1}^{n} (f^i)^T\pi$ is called an F_i-*contrast* , if

$$\sum_{j=1}^{m_i} f^i_j = 0 \ , \quad i \in \{1, \dots, n\} \ , \text{ and}$$

$$f^{i^*} = 0_{m_{i^*}} \quad \text{for all } i^* \in \{1, \dots, n\} \setminus \{i\} \ .$$

F_n-contrasts are also called *treatment contrasts* .

In view of lemma (2.1.10) it is clear that any estimable linear function in the parameters of a sole factor F_i must be an F_i-contrast. Now in particular those designs are important for which *all* F_i-contrasts are estimable (or focusing on the treatment factor, all treatment contrasts are estimable) . This leads to the following definition of the main concept of this work.

2.1.13 Definition

An n-factor design D is called *F_i-connected* if all F_i-contrasts are estimable. F_n-connected designs (all treatment contrasts estimable) are simply called *connected* .

The notion of connectivity was introduced for block designs by Bose [1947] and has in the meantime been studied for general n-factor designs by several authors (cf. e. g. Weeks/Williams [1964], Srivastava/Anderson [1970], Eccleston/Russell [1975]).

For several reasons research on connectivity plays an important role in the theory of experimental designs. It is easy to see that for a given index i all F_i-contrasts are estimable if and only if all *elementary* F_i-contrasts

$$\pi^i_j - \pi^i_{j^*} \quad (j,j^* \in \{1,\dots,m_i\} \ , \ \ j \neq j^*)$$

are estimable (each F_i-contrast can be written as a linear combination of elementary F_i-contrasts). Individual parameters are not estimable as mentioned before (lack of uniqueness), but estimability of all elementary F_i-contrasts is guaranteed for F_i-connected n-factor designs and only for these Furthermore , there is a close connection with the theory of testing linear hypotheses. According to Graybill [1961] the hypothesis

$$\pi^i_1 = \pi^i_2 = \dots = \pi^i_{m_i}$$

is testable if and only if all elementary F_i-contrasts are estimable.

2.2 Graphs

This paragraph contains a short summary of definitions and notations from the theory of graphs which will be useful for stating and proving many of the results to follow.

A *graph* (more precisely: a finite undirected graph) is an ordered pair $G = [V,E]$, where V is a finite, nonempty set and E a finite family of unordered pairs $\{v_1,v_2\}$ such that $v_1,v_2 \in V$. Elements of V are called *vertices* , elements of E *edges* . An edge $e = \{v_1,v_2\} \in E$ is called *incident* to v_1 and v_2 , the vertices v_1 and v_2 are called *adjacent* . $v \in V$ is an *isolated vertex* if no other vertex in V is adjacent to v . $G = [V,E]$ is called a *complete graph* if any two distinct vertices in V are adjacent. A *chain* τ from v_1 to v_t in $G = [V,E]$ is a sequence of vertices in V and edges in E

$$\tau = [v_1,e_1,v_2,e_2,\dots,e_{t-1},v_t]$$

with the property that $e_i = \{v_i,v_{i+1}\}$ for $i = 1,\dots,t-1$. v_1 is called the *tail* vertex and v_t the *head* vertex of the chain τ . A *cycle* is a closed chain, i. e. a chain where head and tail are the same vertex: $v_1 = v_t$. If the vertices $v_1,\dots,v_{t-1}$ of a cycle σ are distinct, σ is an *elementary cycle* .

A graph $G = [V,E]$ is called a *connected graph* if any two vertices $v,v^* \in V$ are linked by a chain. $G^* = [V^*,E^*]$ is a *subgraph* of $G = [V,E]$

if $V^* \subseteq V$ and $E^* \subseteq E$; in case $V^* = V$ and $E^* \neq E$ a subgraph of G is also called *partial graph* of G . A *connected component* of G is a connected subgraph of G which is maximal with respect to the property of being connected. The number of connected components of a graph G is denoted by $c(G)$.

A *tree* is a connected graph which does not contain a cycle. A tree $G^* = [V^*,E^*]$ is called a *spanning tree* of a graph $G = [V,E]$ if $V = V^*$ and $E^* \subseteq E$.

$G = [V,E]$ is a *bipartite graph* if the set of vertices V can be partitioned into two nonvoid subsets V_1 and V_2 (i. e. $V_1 \cup V_2 = V$, $V_1 \cap V_2 = \emptyset$) , such that each edge in E is incident to one vertex in V_1 and one vertex in V_2.

A *digraph* (more precisely: a finite directed graph) $G = (V,E)$ consists of a finite set of vertices V and a finite family E of ordered pairs (v_1,v_2) with $v_1,v_2 \in V$. Elements of E are called *arcs* , v_1 is the *tail* vertex and v_2 the *head* vertex of the arc $e = (v_1,v_2)$, e is *directed from* v_1 *to* v_2 . A *loop* is an arc in which tail and head vertex are identical.

Each digraph $G = (V,E)$ obviously gives rise to an undirected graph $\overline{G} = [V,\overline{E}]$ with the same set of vertices by simply ignoring the orientation of the arcs $(v_1,v_2) \in E$ and taking $\overline{E}$ to be the family of all unordered pairs $\{v_1,v_2\}$ for which $(v_1,v_2) \in E$. Accordingly all notions introduced for graphs (incidence, adjacency, chain, cycle, connectedness, tree, etc.) also apply to digraphs.

For a chain $\tau = [v_1, e_1, \dots, e_{t-1}, v_t]$ in a digraph $G = (V,E)$ let

$$E(\tau^+)$$

denote the family of arcs which appear in τ and are oriented in the direction of τ (i. e. those arcs e_s in τ which are directed from v_s to v_{s+1}, $1 \leq s < t$) . The family of the remaining arcs in τ is denoted by $E(\tau^-)$ (i. e. arcs e_s in τ with $e_s = (v_{s+1}, v_s)$, $1 \leq s < t$) . Finally, a cycle σ in a digraph is called *balanced* if

$$|E(\sigma^+)| = |E(\sigma^-)| ,$$

otherwise *unbalanced* .

Multiple edges between two vertices of a graph and multiple arcs between two vertices of a digraph with the same orientation may occur, for this reason we have spoken of families of edges and arcs instead of sets. Nevertheless set notation will be used for representing families of edges and arcs: possible multiple elements will then be distinguished by means of indices.

2.3. The state of the art

For a two-factor design $B^T B$ reduces to

$$B^T B = \begin{bmatrix} D_1 & D \\ D^T & D_2 \end{bmatrix} .$$

Using the submatrices appearing in this representation we define

$$C^{(1)} = D_2 - D\, D_1^{-1} D^T .$$

It is easy to see that $C^{(1)}$ is a symmetric matrix with row and column sums equal to zero, and that

$$\operatorname{rank} C^{(1)} \leqq m_2 - 1 .$$

By means of $C^{(1)}$ a necessary and sufficient condition for a block design to be connected can be stated as follows:

2.3.1 Theorem (Chakrabarti [1962])

A block design given by a non-negative integer $m_1 \times m_2$ matrix D is connected if and only if

$$\operatorname{rank} C^{(1)} = m_2 - 1 .$$

In order to formulate a further characterization of connected block designs D we consider a bipartite graph which is uniquely determined by D. Let

$$G = [V,E]$$

be defined by

$$V_1 = \{v_1^1,\dots,v_{m_1}^1\} \quad ,$$

$$V_2 = \{v_1^2,\dots,v_{m_2}^2\} \quad ,$$

$$V = V_1 \cup V_2$$

and

$$E = \{\{v_i^1,v_j^2\} \mid d_{ij} > 0,\ v_i^1 \in V_1\ ,\ v_j^2 \in V_2\} \quad .$$

In G there is exactly one vertex for each blocking level and each treatment, and each edge is incident to one vertex corresponding to a blocking level and one vertex corresponding to a treatment. v_i^1 is adjacent to v_j^2 if and only if the i-th blocking level occurs in at least one experiment involving the j-th treatment. This representation turns out to be in agreement with the definition of connectedness introduced by Bose [1947] (Bose defined two treatments j , j^* to be connected if v_j^2 and $v_{j^*}^2$ are linked by a chain in G . He called a block design connected if any two treatments are connected) .

For the following theorem we present a new proof.

2.3.2 Theorem (Bose and Gaffke, cf. Krafft [1978] and Gaffke [1978])

A block design D is connected if and only if the bipartite graph G given by D is connected.

Proof:

(i) D is a connected block design if and only if

$$\forall \ \{v_i^1, v_j^2\} \in E \quad \pi_i^1 + \pi_j^2 = 0 \tag{2.3.3}$$

$$\Longrightarrow \exists \ \lambda \in \mathbb{R} \quad \pi_2 = \lambda \cdot 1_{m_2} \quad .$$

This holds because by lemma (2.1.9) connectivity of D is equivalent to

$$B \ \pi = 0 \quad \Longrightarrow$$

$$(f^2)^T \ \pi^2 = 0 \quad \forall \ f^2 \in \mathbb{R}^{m_2} \ \text{ with } \sum_{j=1}^{m_2} f_j^2 = 0$$

and because

$$B \ \pi = 0 \quad \Longleftrightarrow \quad \pi_i^1 + \pi_j^2 = 0 \quad \forall \ i,j \ \text{ for which } \ d_{ij} > 0$$

$$\Longleftrightarrow \quad \pi_i^1 + \pi_j^2 = 0 \quad \forall \ \{v_i^1, v_j^2\} \in E$$

as well as

$$\exists \ \lambda \in \mathbb{R} \ \ \pi^2 = \lambda \cdot 1_{m_2} \quad \Longleftrightarrow \quad (f^2)^T \ \pi^2 = 0 \qquad \forall \ f^2 \in \mathbb{R}^{m_2}$$

$$\text{for which } \sum_{j=1}^{m_2} f_j^2 = 0 \quad .$$

(ii) G is a connected graph if and only if the implication (2.3.3) holds.

"$\Longrightarrow$": Any two vertices $v_j^2, v_{j*}^2 \in V_2$ are linked by a chain in G if G is a connected graph. Hence the premise of (2.3.3) implies $\pi_j^2 = \pi_{j*}^2$.

"$\Longleftarrow$": If G is not connected, one can find vertices $v_j^2, v_{j*}^2 \in V_2$ which are not linked by a chain (by (2.1.4) it follows that there are no isolated vertices in V_1). Let $G_j = [\tilde{V}_j, E_j]$ be the connected component of G containing v_j^2 . Define

$$\hat{\pi}_j^2 = \begin{cases} 1 & \text{if } v_j^2 \in V_2 \cap \tilde{V}_j , \\ 0 & \text{otherwise} , \end{cases}$$

$$\hat{\pi}_i^1 = \begin{cases} -1 & \text{if } v_i^1 \in V_1 \cap \tilde{V}_j , \\ 0 & \text{otherwise} , \end{cases}$$

where $i = 1,\dots,m_1$, $j = 1,\dots,m_2$. It follows that $B\,\hat{\pi} = 0$ but

$$\hat{\pi}^2 \neq \lambda \cdot 1_{m_2} \quad \forall\, \lambda \in \mathbb{R} ,$$

hence D cannot be connected.

□

2.3.4 Example

Consider the block design given by

$$D = \begin{bmatrix} 2 & 1 & 0 & 0 & 0 \\ 0 & 5 & 3 & 0 & 0 \\ 0 & 0 & 1 & 0 & 0 \\ 0 & 1 & 0 & 2 & 2 \end{bmatrix} .$$

The associated bipartite graph G

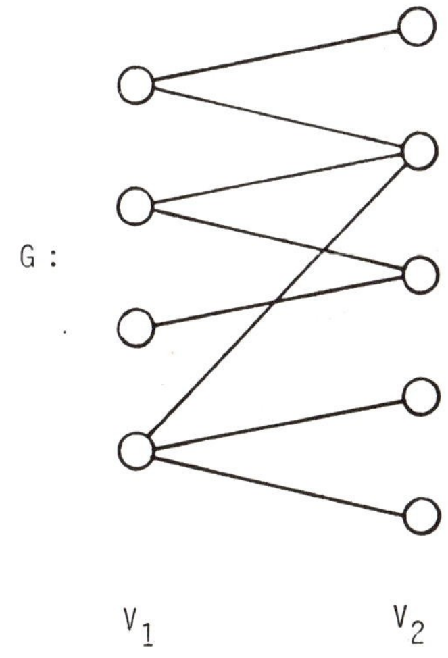

is connected, hence D is a connected design .

2.3.5 Example

Now let D be the design which results from the one in the preceding example by eliminating one experiment, namely the one where the second treatment is applied to the fourth level of the blocking factor, i. e.

$$D = \begin{bmatrix} 2 & 1 & 0 & 0 & 0 \\ 0 & 5 & 3 & 0 & 0 \\ 0 & 0 & 1 & 0 & 0 \\ 0 & 0 & 0 & 2 & 2 \end{bmatrix} .$$

The appropriate graph G is obtained by deleting the edge $\{v_4^1 , v_2^2\}$ in the graph of example (2.3.4) . It follows that the design in question is not connected.

2.3.6 Remark

Clearly a block design is F_1-connected if and only if it is connected (i. e. F_2-connected).

Next we shall report on characterizations of connected designs for two-way elimination of heterogeneity (three-factor designs) which are known from the literature. In analogy to theorem (2.3.1) there is a criterion involving submatrices of

$$B^T B = \begin{bmatrix} D_1 & D_{12} & D_{13} \\ D_{12}^T & D_2 & D_{23} \\ D_{13}^T & D_{23}^T & D_3 \end{bmatrix} .$$

Let X^- denote a generalized inverse of a given matrix X (not necessarily the Moore-Penrose inverse; X^- may be any matrix satisfying $X X^- X = X$) . We define

$$\hat{D} = D_{23}^T - D_{13}^T D_1^{-1} D_{12} \quad ,$$

$$C^{(2)} = D_3 - D_{13} D_1^{-1} D_{13}^T - \hat{D} (D_2 - D_{12}^T D_1^{-1} D_{12})^- \hat{D}^T \; .$$

2.3.7 Theorem (Raghavarao/Federer [1975])

Let $C^{(2)}$ be defined for a design for two-way elimination of heterogeneity. The design is connected if and only if

$$\text{rank } C^{(2)} = m_3 - 1 \; .$$

In general this criterion cannot be checked easily, therefore a great deal of research has been devoted to the investigation of special cases. Row-column designs (cf. section 2.1) are an important subclass of three-factor designs.

Consider a row-column design D which is given by an $m_1 \times m_2$ matrix Y where

$$d_{ijk} = \begin{cases} 1 & \text{if } y_{ij} = k \, , \\ 0 & \text{otherwise} \, . \end{cases}$$

It is easy to verify that

$$B^T B = \begin{bmatrix} m_2 \cdot I_{m_1} & 1_{m_1,m_2} & D_{13} \\ 1_{m_2,m_1} & m_1 \cdot I_{m_2} & D_{23} \\ D_{13}^T & D_{23}^T & D_3 \end{bmatrix} \quad ,$$

where I_m denotes the $m \times m$ identity matrix and $1_{m,m^*}$ an $m \times m^*$ matrix of all ones. Defining

$$C_{rc}^{(2)} = D_3 - m_2^{-1} D_{13}^T D_{13} - m_1^{-1} D_{23}^T D_{23} - (m_1 m_2)^{-1} D_3 1_{m_3,m_3} D_3$$

one obtains a result which is in perfect analogy to theorem (2.3.7) .

<u>2.3.8 Theorem</u> (Chakrabarti [1962])

Let $C_{rc}^{(2)}$ be defined for a row-column design. The design is connected if and only if

$$\text{rank } C_{rc}^{(2)} = m_3 - 1 \quad .$$

Wynn [1977] gave a combinatorial characterization of connected row-column designs with only $m_2 = 2$ columns. Consider the following digraph defined by an $m_1 \times 2$ matrix Y where y_{i1} , $y_{i2} \in \{1,\dots,m_3\}$ for $i = 1,\dots,m_2$:

$$G = (V,E) \quad ,$$

$$V = \{1,\dots,m_3\} \quad ,$$

$$E = \{(y_{i2},y_{i1}) \mid 1 \leq i \leq m_1\} \quad .$$

Wynn proved the following result by means of network flow theory. We present a shorter proof which makes use of a different approach.

2.3.9 Theorem (Wynn [1977])

Let $\{1,\ldots,m_3\}$ be the set of entries of an $m_1 \times 2$ matrix Y . The row-column design given by Y is connected if and only if the digraph G associated with Y is connected and contains an unbalanced cycle.

Proof:

We apply lemma (2.1.9) .

(i) Let $k \in V$ be assigned to the treatment parameters π_k^3 , $k = 1,\ldots,m_3$. Eliminating the components of π^1 in the system $B\pi = 0$ one obtains

$$\exists\, \pi^1 \in \mathbb{R}^{m_1} \quad B \begin{pmatrix} \pi^1 \\ \pi^2 \\ \pi^3 \end{pmatrix} = 0 \iff$$

$$\forall\, e_i = (k,k^*) \in E \quad \pi_{k^*}^3 - \pi_k^3 = \pi_2^2 - \pi_1^2 \,.$$

Consider an arbitrary $\hat{\pi}$ in the kernel of B and let

$$\delta = \hat{\pi}_2^2 - \hat{\pi}_1^2 \quad .$$

For a chain τ from k to k' in G it follows that

$$\hat{\pi}_{k'}^3 - \hat{\pi}_k^3 = \sum_{e \in E(\tau^+)} \delta \; - \sum_{e \in E(\tau^-)} \delta \;.$$

Now let σ be a cycle in G with $|E(\tau^+)| \neq |E(\tau^-)|$ and $k \in V$ a vertex contained in this cycle. Hence

$$\hat{\pi}_k^3 - \hat{\pi}_k^3 = 0 \;=\; (|E(\sigma^+)| - |E(\sigma^-)|) \cdot \delta$$

i. e. $\delta = 0$. In the case of a connected digraph G we conclude for $\hat{\pi}^3$:

$$\hat{\pi}^3 = \lambda \cdot 1_{m_3} \quad , \quad \lambda \in \mathbb{R} \; .$$

As a consequence the design has to be connected.

(ii) If G is not connected, consider a connected component $G^* = (V^*,E^*)$ of G and let

$$\hat{\pi}^3_k = \begin{cases} 0 & \text{for } k \in V^* , \\ 1 & \text{for } k \in V \setminus V^* \end{cases} ,$$

$$\hat{\pi}^2_1 = \hat{\pi}^2_2 = 0 \; ,$$

$$\hat{\pi}^1_i = -\hat{\pi}^3_{y_{i1}} \; .$$

For the resulting vector $\hat{\pi}$ we have $B\,\hat{\pi} = 0$ but $\hat{\pi}^3 \neq \lambda \cdot 1_{m_3}$ for all $\lambda \in \mathbb{R}$.

If G is a connected digraph which has no unbalanced cycle, consider a parameter vector $\hat{\pi}$ with

$$\hat{\pi}^2_1 = 0 \quad , \quad \hat{\pi}^2_2 = \delta > 0 \quad ,$$

$$\hat{\pi}^3_{y_{11}} = 0 \quad , \quad \hat{\pi}^3_{y_{12}} = -\,\delta \quad ,$$

$$\hat{\pi}^1 = 0_{m_1} \quad .$$

This can easily be extended to a solution of $B\pi = 0$:

Let τ_k be chain from $k^* = y_{11}$ to k, then we take

$$\hat{\pi}_k^3 = (|E(\tau_k^+)| - |E(\tau_k^-)|)\cdot\delta \quad , \quad 1 \leq k \leq m_3 \ .$$

This proves the theorem as $\hat{\pi}^3_{y_{11}} = \hat{\pi}^3_{y_{12}}$. □

Extending the result of Wynn [1977] this proof technique yields:

2.3.10 Corollary

A row-column design with only two columns is connected if and only if it is F_1-connected and F_2-connected.

Proof:

According to the first part of the proof of theorem (2.3.9), the existence of an unbalanced cycle implies that for a vector $\hat{\pi}$ in the kernel of B we have $\hat{\pi}_1^2 = \hat{\pi}_2^2$. Hence connectivity implies F_2-connectivity.

On the other hand connectedness of the associated digraph yields $\hat{\pi}^3 = \lambda \cdot 1_{m_3}$, i. e. together

$$\hat{\pi}^1 = -(\hat{\pi}_1^2 + \hat{\pi}_1^3)\cdot 1_{m_1} \quad ,$$

hence F_1-connectivity is also guaranteed.

Conversely, in the case of a design which is F_1-connected and F_2-connected we have

$$B\,\hat{\pi} = 0 \implies \hat{\pi}^1 = \mu \cdot 1_{m_1} \quad \text{and} \quad \hat{\pi}_1^2 = \hat{\pi}_2^2$$

$$\implies \hat{\pi}^3 = -(\mu + \hat{\pi}_1^2) \cdot 1_{m_3}$$

where $\mu \in \mathbb{R}$, i. e. the design is connected . □

For row-column designs with only two rows one obtains an analogous characterization by considering the digraph associated with Y^T . For general designs for two-way elimination of heterogeneity or even for row-column designs with arbitrarily many rows and columns we are not aware of any combinatorial characterization of connectivity beyond the results presented in the following chapters.

Cheng [1978] showed how to calculate a matrix generalizing the matrices $C^{(1)}$ and $C^{(2)}$ from theorems (2.3.1) and (2.3.7) , resp.,for n-factor designs D with

$$\sum_{j_n=1}^{m_n} d_{j_1 \dots j_n} = t \quad ,$$

where t is a given integer. Gateley [1962] (unpublished paper, quoted in Weeks/Williams [1964]) gave a criterion for the estimability of all F_i-contrasts , $i = 1, \dots, n$, in an n-factor design .

<u>2.3.11 Definition</u>

An n-factor design D is called *completely connected* , if D is F_i-connected for all i , $1 \leq i \leq n$.

We had no access to the paper by Gateley [1962], but his result follows readily by the above approach .

2.3.12 Theorem (Gateley [1962])

An n-factor design D is completely connected if and only if the corresponding matrix B satisfies

$$\text{rank } B = \sum_{i=1}^{n} m_i - n + 1 \quad .$$

Proof:

By lemma (2.1.9) , D is completely connected if and only if

$$B\,\pi = 0 \implies \pi^i = \lambda_i \cdot 1_{m_i} \quad , \quad 1 \leq i \leq n \quad , \quad \lambda_i \in \mathbb{R}$$

$$\iff \quad \text{kernel } B = \left\{ \pi \mid \pi^i = \lambda_i \cdot 1_{m_i} \;,\; \lambda_i \in \mathbb{R} \;,\; 1 \leq i \leq n \;, \right.$$

$$\left. \pi^n = -\left(\sum_{i=1}^{n-1} \lambda_i\right) \cdot 1_{m_n} \right\}$$

$$\iff \quad \dim \text{kernel } B = n - 1$$

$$\iff \quad \text{rank } B = \sum_{i=1}^{n} m_i - n + 1 \quad . \qquad \square$$

Weeks/Williams [1964] gave a simple method for determining completely connected multi-factor designs: For a given design D each row of the matrix B corresponds to a single experiment of the design (cf. section 2.1). Two experiments are called *nearly identical* if the associated row vectors of B differ in at most one component. Consider for example a graph $G = [V,E]$ that contains exactly one vertex for each experiment of the design (i. e. for every row of B) and in which two vertices are adjacent if and only if their experiments are nearly identical. Now it is easy to see that the method suggested in Weeks/Williams [1964] checks whether G is a connected graph or not. Obviously for any row-column design with $m_3 \geq 2$ treatments the resulting graph G is not connected. In the errata postscript to their paper the authors emphasize the fact that their method yields only a sufficient criterion for a design to be completely connected. In fact it is an easy consequence of lemma (2.1.9) that connectedness of the graph G implies that the underlying design has to be completely connected.

Another characterization of connectivity postulated by Ecclestone [1972] will be discussed at the end of section 3.5 .

In the next chapter we give characterizations of arbitrary connected (but not necessarily completely connected) n-factor designs, and we also show how the combinatorial characterization of connected block designs and row-column designs can be extended to the general case.

3. CHARACTERIZATIONS OF CONNECTED DESIGNS BY MEANS OF DIGRAPHS WITH LABELED CYCLES

3.1 Prerequisites

First we introduce some technical notations. For $j \in \{1,\dots,m_1\}$ let A_j be a lower triangular $d_j^1 \times d_j^1$ matrix, namely

$$A_j = \begin{bmatrix} 1 & & & & \\ 1 & -1 & & 0 & \\ 1 & & -1 & & \\ \vdots & & 0 & \ddots & \\ 1 & & & & -1 \end{bmatrix}.$$

According to (2.1.4) all d_j^1 are positive. Using $A_1,\dots,A_{m_1}$ we define a $p \times p$ matrix A of block diagonal form,

$$A = \begin{bmatrix} A_1 & & & \\ & A_2 & & 0 \\ & & \ddots & \\ & 0 & & \\ & & & A_{m_1} \end{bmatrix}$$

where $p = \sum_{j=1}^{m_1} d_j^1$ (cf. section 2.1).

As mentioned before, the rows of B are assumed to be in an order that agrees with the lexicographical ordering of the index tuples $(j_1,\dots,j_n)$ of their non-zero coordinates. Therefore multiplication of $B\pi = 0$ from the left by A eliminates the parameters π_j^1 $(j = 1,\dots,m_1)$ by means of appropriate row subtractions. In order to rearrange the columns of A B , consider the following representation of the $p \times p$ identity matrix:

$$I_p = \Big(u^1, u^2, \dots, u^{s_1}, u^{s_1+1}, \dots, u^{s_2}, \dots, u^{s_i}, u^{s_i+1}, \dots, u^{s_{i+1}}, \dots,$$
$$u^{s_{m_1-1}}, u^{s_{m_1-1}+1}, \dots, u^{s_{m1}}\Big)$$

where

$$s_i = \sum_{j=1}^{i} d_j^1 \quad , \qquad i = 1,\dots,m_1 \quad ,$$

u^k is the k-th unit vector of $\mathbb{R}^p$. Clearly $p = s_{m_1}$.

Reordering the column vectors of I_p in a certain way one obtains the following $p \times p$ permutation matrix:

$$P = \Big(u^1, u^{s_1+1}, u^{s_2+1}, \dots, u^{s_{m_1-1}+1},$$
$$u^2, u^3, \dots, u^{s_1}, u^{s_1+2}, u^{s_1+3}, \dots, u^{s_2}, \dots,$$
$$u^{s_i+2}, u^{s_i+3}, \dots, u^{s_{i+1}}, \dots,$$
$$u^{s_{m_1-1}+2}, u^{s_{m_1-1}+3}, \dots, u^{s_{m_1}}\Big)$$

We have chosen P in such a way that multiplying A B from the left by P results in a matrix of the form

$$P\,A\,B = \begin{pmatrix} I_{m_1} & \tilde{B}_2 & \tilde{B}_3 & & \tilde{B}_n \\ 0_{p-m_1,m_1} & B_2 & B_3 & \cdots & B_n \end{pmatrix} .$$

Here the first block of rows $(I_{m_1}, \tilde{B}_2, \tilde{B}_3, \ldots, \tilde{B}_n)$ is a submatrix of B. $0_{m,m^*}$ denotes an $m \times m^*$ matrix of all zeros. Each row of B_i $(i = 2,\ldots,n)$ is the difference of two unit vectors in $\mathbb{R}^{m_i}$.

Now we can give a necessary and sufficient characterization of connectivity for arbitrary multi-factor designs.

<u>3.1.2 Lemma</u>

An n-factor design given by the matrix B of the model (2.1.1) is connected if and only if

$$\text{rank } B = m_1 + m_n - 1 + \text{rank } (B_2,\ldots,B_{n-1}) .$$

<u>Proof:</u>

The matrices P and A in (3.1.1) are non-singular, therefore

$$\text{kernel } B = \text{kernel } P\,A\,B .$$

The representation of $P\,A\,B$ in (3.1.1) implies

$$\text{kernel } B = \{ \pi \mid ((\pi^2)^T,\ldots,(\pi^n)^T)^T \in \text{kernel } (B_2,\ldots,B_n),$$

$$\pi^1 = - \sum_{i=2}^{n} \tilde{B}_i \, \pi^i \} .$$

As a consequence of lemma (2.1.9) connectivity holds if and only if for each $\hat{\pi} \in$ kernel B there exists a $\lambda \in \mathbb{R}$ such that $\hat{\pi}^n = \lambda \cdot 1_{m_n}$. Consider the following set $X \subseteq \mathbb{R}_q$:

$$X = \{ \pi | ((\pi^2)^T, \dots, (\pi^{n-1})^T)^T \in \text{kernel } (B_2, \dots, B_{n-1}) ,$$

$$\exists \lambda \in \mathbb{R} \text{ with } \pi^n = \lambda \cdot 1_{m_n} , \pi^1 = - \sum_{i=2}^{n} \tilde{B}_i \pi^i \} .$$

Clearly

$$X \subseteq \text{kernel } B ,$$

$$\dim X = \text{rank } (B_2, \dots, B_{n-1}) + 1 ,$$

therefore

$$\text{rank } B \leq m_1 + m_n - 1 + \text{rank } (B_2, \dots, B_{n-1}) .$$

The design is connected if and only if $X =$ kernel B , which completes the proof.

□

For the next lemma we associate graphs with matrices which have the same form as $B_2, \dots, B_n$. Consider an $s \times m$ matrix Q in which each row q^i can be represented as the difference of two m-dimensional unit vectors. The associated graph G_Q is defined by

$$G_Q = [V,E] ,$$

$$V = \{1, \dots, m\} ,$$

$$E = \{\{k,k^*\} \mid \exists\ i,\ 1 \leq i \leq s,\ (q^i)^T = u^{k^*} - u^k \neq 0_m\}\ ,$$

where u^k and u^{k^*} denote the k-th and k^*-th unit vector of $\mathbb{R}^m$ respectively.

<u>3.1.3 Lemma</u>

G_Q has exactly m - rank Q connected components, i. e.

$$\text{rank } Q = m - c \iff c = c(G_Q)\ .$$

<u>Proof:</u>

Let us exclude the trivial case $Q = 0_{s,m}$. Then we can without loss of generality suppose that the rows and columns of Q have been arranged as follows:

$Q =$

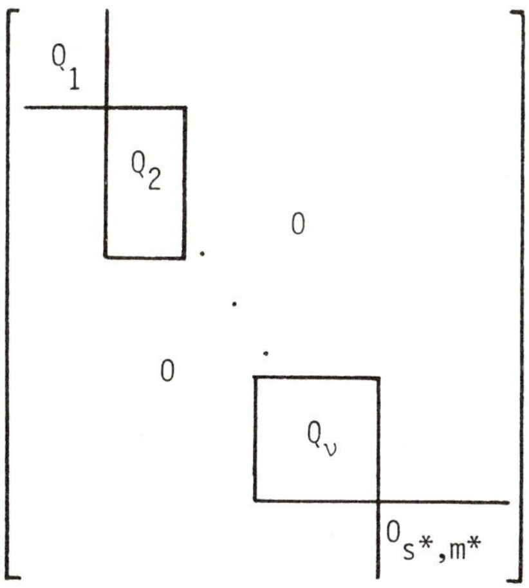

,

where the set of column indices of each Q_j corresponds to the vertex set V_j of a nontrivial connected component $(|V_j| > 1)$

$$G_{Q_j} = [V_j, E_j]$$

of G_Q . Let Q_j be $s_j \times m_j$ -matrices (clearly they have no columns of all zeros), then

$$s = \sum_{j=1}^{\nu} s_j + s^* ,$$

$$m = \sum_{j=1}^{\nu} m_j + m^* ,$$

where $\nu \in \mathbb{N}$, $s^*, m^* \in \mathbb{N}_0$ ($\mathbb{N}$ and $\mathbb{N}_0$ denote the set of positive and nonnegative integer numbers, respectively). It follows that

$$\nu = c(G_Q) - m^* ,$$

where m^* is the number of isolated vertices in the graph G_Q. It is easy to see that two vertices k, k^* belong to the same connected component G_{Q_j} if and only if for all $z \in \text{kernel } Q_j$ the corresponding components z_k and z_{k^*} are the same. Therefore

$$\text{rank } Q_j = |V_j| - 1$$

and

$$\begin{aligned} \text{rank } Q &= m - m^* - \nu \\ &= m - c(G_Q) . \end{aligned}$$ □

Now in addition let a real $s \times t$ matrix R be given and define a digraph

$$G_{R,Q} = (V,\tilde{E})$$

where

$$V = \{1,\dots,m\} \quad ,$$

$$\tilde{E} = \{(k,k\ )_r \mid \exists \text{ row } (r,q) \text{ of } (R,Q) \text{ with } q^T = u^{k^*} - u^k \ ,\ k,k^* \in V\} \ .$$

By construction it follows that for a row of (R,Q) of the form $(r,0_m^T)$, the digraph contains a loop $(k,k)_r$ at each vertex $k \in V$. The vector "indices" r of the arcs are used to define a labeling function $h: \tilde{E} \longrightarrow \mathbb{R}^t$,

$$h(e) = r^T \quad \text{for} \quad e = (k,k^*)_r \in \tilde{E} \quad ,$$

which assigns to each arc the transpose of the corresponding row of R. In a straightforward way the domain of h can be extended to the set Θ of all chains τ in $G_{R,Q}$, namely by defining

$$h(\tau) = \sum_{e \in E(\tau^+)} h(e) - \sum_{e \in E(\tau^-)} h(e) \quad , \quad \tau \in \Theta \ .$$

3.1.4 Definition

Let $h: \Theta \longrightarrow \mathbb{R}^t$ be a function assigning a t-vector $h(\tau)$ to each chain τ of a given digraph. A set of ℓ cycles $\sigma_1,\dots,\sigma_\ell \in \Theta$ is called h-*independent* if the vectors $h(\sigma_1),\dots,h(\sigma_\ell)$ are linearly independent in $\mathbb{R}^t$. We simply refer to an *independent* set of vector labels if there is no risk of confusion with respect to the underlying labeling funktion h .

The following two lemmas will turn out to be useful tools for proving several results in the remaining sections of this chapter.

3.1.5 Lemma

Consider a real $s \times t$ matrix R and an $s \times m$ matrix Q with the property that each row of Q is the difference of two unit vectors. Let $G_{R,Q}$ be the associated digraph with a labeling function $h: \Theta \longrightarrow \mathbb{R}^t$ as described above. Then the following holds:

$$\text{rank } (R,Q) = \text{rank } Q + \ell$$

$$\Longleftrightarrow \quad \text{there are exactly } \ell \text{ independent cycles in } G_{R,Q} \ .$$

Proof:

Let $z_e = (r_e, q_e)$ denote the row of (R,Q) which corresponds to the arc $e \subset \tilde{E}$, and let (R_E, Q_E) be the submatrix corresponding to a set of arcs $E \subseteq \tilde{E}$. Choose $E' \subseteq \tilde{E}$ with $|E'| = \text{rank } Q$ such that

$$\text{rank } Q = \text{rank } Q_{E'} \ .$$

As a consequence E' is "cycle-free", i. e. there is no cycle in $G_{R,Q}$ with the property that all of its arcs are contained in the set E' . By lemma

(3.1.3) we have

$$\text{rank } Q = m - c(G_{R,Q})$$

which - according to an elementary theorem from graph theory - is the maximum cardinality of a cycle-free set of arcs in a digraph having m vertices and $c(G_{R,Q})$ connected components.

(i) Consider $E'' \subseteq \tilde{E} \setminus E'$ with $|E''| = \ell$ and

$$\text{rank} \begin{pmatrix} R_{E'} & Q_{E'} \\ R_{E''} & Q_{E''} \end{pmatrix} = |E'| + |E''| = \text{rank } Q + \ell \quad .$$

The existence of such a set E'' is guaranteed provided that rank $(R,Q) = \text{rank } Q + \ell$. Choose any

$$\tilde{e} = (k,k^*)_r \in E''$$

and denote the corresponding row vector of (R,Q) by

$$z_{\tilde{e}} = (r,(u^{k^*} - u^k)^T) \quad .$$

Because of $\tilde{e} \notin E'$ there exists a chain τ in $G_{R,Q}$ from k^* to k such that each arc of τ belongs to the set E' - otherwise E' could not be a maximal cycle-free set of arcs.

Now

$$(h(\tau)^T, (u^k - u^{k^*})^T) = \sum_{e \in E(\tau^+)} (r_e, q_e) - \sum_{e \in E(\tau^-)} (r_e, q_e) .$$

In case $k^* = k$ we have $\tau = [k]$ and $h(\tau) = 0_t$. Together $\tilde{e}$ and τ form an elementary cycle $\sigma_{\tilde{e}}$ in $(V, E' \cup \{\tilde{e}\})$,

$$(h(\sigma_{\tilde{e}})^T, 0_m^T) = (r, (u^{k^*} - u^k)^T) + (h(\tau)^T, (u^k - u^{k^*})^T) .$$

Now let $H_{E''}$ be an $\ell \times t$ matrix having as rows the vectors $h(\sigma_e)^T$, $e \in E''$. By construction

$$(H_{E''}, 0_{\ell,m}) = (R_{E''}, Q_{E''}) + M \cdot (R_{E'}, Q_{E'}) ,$$

where M is an $|E''| \times |E'|$ matrix with entries in the set $\{-1,0,1\}$, i. e. the matrix

$$\begin{pmatrix} R_{E'} & Q_{E'} \\ H_{E''} & 0_{\ell,m} \end{pmatrix}$$

can be obtained by elementary row operations from the matrix

$$\begin{pmatrix} R_{E'} & Q_{E'} \\ R_{E''} & Q_{E''} \end{pmatrix} .$$

Hence

$$\operatorname{rank}\begin{pmatrix} R_{E'} & Q_{E'} \\ H_{E''} & 0_{\ell,m} \end{pmatrix} = \operatorname{rank}\begin{pmatrix} R_{E'} & Q_{E'} \\ R_{E''} & Q_{E''} \end{pmatrix} ,$$

$$\operatorname{rank} Q_{E'} + \operatorname{rank} H_{E''} = \operatorname{rank} Q + \ell ,$$

$$\operatorname{rank} H_{E''} = \ell ,$$

consequently there have to be at least ℓ independent cycles in $G_{R,Q}$.

(ii) To prove the converse, let H denote the $\ell \times t$ matrix H with rows $h(\sigma_1)^T,\dots,h(\sigma_\ell)^T$. There exists an $\ell \times s$ matrix M^* such that

$$(H,0_{\ell,m}) = M^* \cdot (R,Q) .$$

It follows that

$$\operatorname{rank}(R,Q) \geqq \operatorname{rank}\begin{pmatrix} R_{E'} & Q_{E'} \\ H & 0_{\ell,m} \end{pmatrix}.$$

$$= \operatorname{rank} Q + \ell ,$$

which completes the proof.

□

3.1.6 Lemma

The hypotheses of lemma (3.1.5) also imply the following equivalence:

$$\text{rank } (R,Q) = \text{rank } R + m - 1$$

$\Longleftrightarrow$ $G_{Q,R}$ is a connected digraph containing exactly $\ell = \text{rank } R$ independent cycles .

Proof:

The rank of Q is at most $m - 1$ as $Q \cdot 1_m = 0_s$. Hence

$$\begin{aligned} \text{rank } R + m - 1 &= \text{rank } (R,Q) \\ &\leq \text{rank } R + \text{rank } Q \\ &\leq \text{rank } R + m - 1 \quad , \end{aligned}$$

and this implies

$$\text{rank } Q = m - 1$$

which, according to lemma (3.1.3) holds if and only if the digraph $G_{Q,R}$ is connected. The rest follows from the proof of lemma (3.1.5) . □

It is easy to verify that lemmas (3.1.5) and (3.1.6) remain true after replacing "cycles" by "elementary cycles" .

3.2 Row-column designs

In this section we consider row-column designs, i. e. designs for two-way elimination of heterogeneity (three-factor designs) given by a nonnegative integer $m_1 \times m_2 \times m_3$ matrix D with the property that

$$D_{12} = 1_{m_1,m_2} \quad .$$

As mentioned in chapter 2 , such designs can be represented by $m_1 \times m_2$ matrices Y , where $y_{ij} \in \{1,\dots,m_3\}$,

$$y_{ij} = k \quad \Longleftrightarrow \quad d_{ijk} > 0 \quad .$$

With each row-column design we now associate a uniquely defined digraph G_{rc} whose arcs are labeled with integer numbers:

$$G_{rc} = (V,E) \quad ,$$

$$V = \{1,\dots,m_3\} \; ,$$

$$E = \{(y_{ij},y_{i1})_j \mid 1 \leq i \leq m_1 \; , \; 2 \leq j \leq m_2\} \quad .$$

For the i-th row of Y there are $m_2 - 1$ arcs $(k,k^*)_j$ in E , $j = 2,\dots,m_2$, whose common head vertex is $k^* = y_{i1}$ and whose tail vertices correspond to the remaining entries $y_{i2},\dots,y_{im_2}$ of the i-th row of Y . The column indices j of the tail vertices $k = y_{ij}$ define the labels of the arcs; this is denoted by a subindex j .

3.2.1 Example

Take $m_1 = 5$, $m_2 = 3$, $m_3 = 5$, and let a row-column design be given by means of

$$Y = \begin{bmatrix} 2 & 1 & 4 \\ 5 & 4 & 3 \\ 3 & 3 & 4 \\ 1 & 5 & 3 \\ 1 & 5 & 5 \end{bmatrix} .$$

The following figure shows the digraph G_{rc} and also the labeling of its arcs.

G_{rc} :

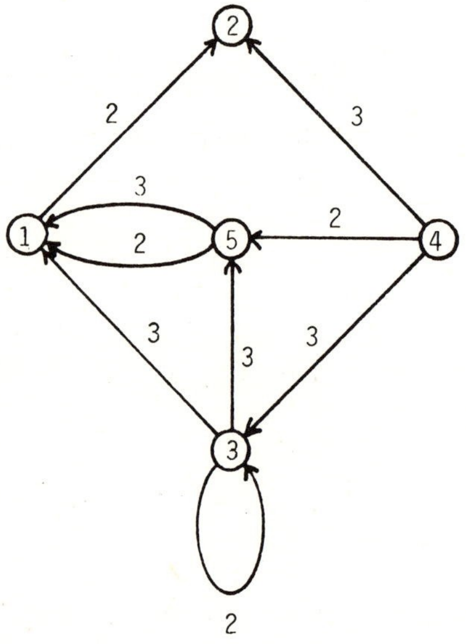

The arcs $(1,2)_2$ and $(4,2)_3$ correspond to the first row of Y , $(4,5)_2$ and $(3,5)_3$ to the second row , etc.

Multiple arcs between two vertices occur only in case of different orientations and / or different labels as indicated by the set notation of E .

We now allocate vector labels to all chains in G_{rc} by means of the labeling of the arcs. Let $e_j \in E$ be an arc with label j, let Θ denote the set of all chains in G_{rc} and let u^j be the j-th unit vector of $\mathbb{R}^{m^2}$. Define a labeling function

$$g : \Theta \longrightarrow \mathbb{Z}^{m^2}$$

(where $\mathbb{Z}^m$ denotes the set of integer m-vectors) by

$$g(\tau) = \sum_{e_j \in E(\tau^+)} u^j - \sum_{e_j \in E(\tau^-)} u^j \quad , \quad \tau \in \Theta .$$

Consider the following chains which occur in the digraph G_{rc} of example (3.2.1) :

τ_1 :

τ_2 :

τ_3 :

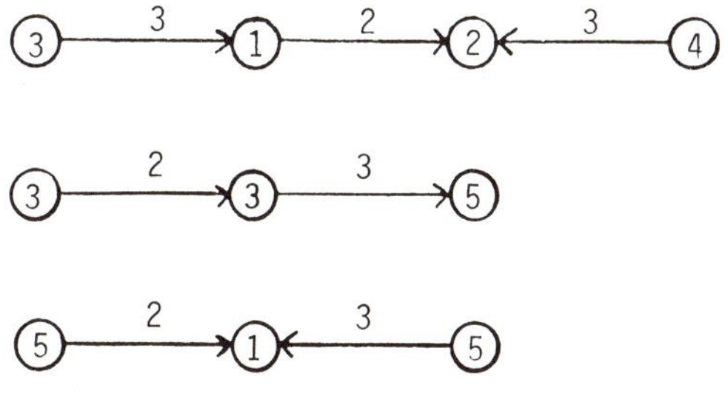

Their respective vector labels are

$$g(\tau_1) = (0\,,1\,,0)^T \quad ,$$

$$g(\tau_2) = (0\,,1\,,1)^T \quad ,$$

$$g(\tau_3) = (0\,,1\,,-1)^T \quad .$$

The following theorem gives a necessary and sufficient characterization of connected row-column designs in terms of the number of cycles in G_{rc} with linearly independent vector labels (cf. Butz [1980]) . The labels occurring in E are $2,\dots,m_2$, therefore any digraph G_{rc} has at most $m_2 - 1$ linearly independent vector labels.

3.2.2 Theorem

Consider a row-column design defined by an $m_1 \times m_2$ matrix Y with $y_{ij} \in \{1,\dots,m_3\}$. The design is connected if and only if its associated digraph G_{rc} is connected and contains $m_2 - 1$ g-independent cycles.

Proof:

Let B be the matrix of the model (2.1.1) for the design in question. According to (3.1.1) , elementary row operations yield

$$P\,A\,B = \begin{pmatrix} I_{m_1} & \tilde{B}_2 & \tilde{B}_3 \\ 0_{p-m_1,m_1} & B_2 & B_3 \end{pmatrix} ,$$

and by lemma (3.1.2) we know that the design is connected if and only if

$$\text{rank}\ (B_2,B_3) = m_3 - 1 + \text{rank}\ B_2$$

For a row-column design it follows by construction of PAB that

$$B_2 = \begin{bmatrix} M \\ \cdot \\ \vdots \\ M \end{bmatrix} \left.\vphantom{\begin{bmatrix} M \\ \cdot \\ \vdots \\ M \end{bmatrix}}\right\} m_1 \text{ times} ,$$

where M is an $(m_2 - 1) \times m_2$ matrix,

$$M = \begin{bmatrix} 1 & -1 & & & & \\ 1 & & -1 & & 0 & \\ & & & \cdot & & \\ & & 0 & & \cdot & \\ & & & & & \cdot \\ 1 & & & & & -1 \end{bmatrix} .$$

As a consequence,

$$\begin{aligned} \text{rank } B_2 &= \text{rank } M \\ &= m_2 - 1 . \end{aligned}$$

Application of lemma (3.1.6) yields that the design is connected if and only if the digraph G_{B_2,B_3} is connected and contains $m_2 - 1$ h-independent cycles $(R = B_2 , Q = B_3 , s = m_1 \cdot (m_2 - 1) , m = m_3 , t = m_2)$. Now let us compare the digraphs $G_{rc} = (V,E)$ and $G_{B_2,B_3} = (V,\tilde{E})$. Denoting $r = (u^1 - u^j)^T$ we get

$$\exists \ k \in V \ \text{ such that } \ (k,k)_r \in \tilde{E}$$

$$\Longleftrightarrow$$

$$\exists \ k^* \in V \ \text{ such that } \ (k^*,k^*)_j \in E$$

and for $k,k^* \in V$, $k \neq k^*$,

$$(k,k^*)_r \in \tilde{E} \quad \Longleftrightarrow \quad (k,k^*)_j \in E \ .$$

This shows how one can associate with each elementary cycle σ in G_{rc} in an obvious way an elementary cycle $\tilde{\sigma}$ in G_{B_2,B_3} and vice versa such that

$$h(\tilde{\sigma}) = - g(\sigma) + (|E(\sigma^+)| - |E(\sigma^-)|) \ u^1 \ .$$

Observing that

$$(|E(\sigma^+)| - |E(\sigma^-)|) = g(\sigma)^T \cdot 1_{m_2}$$

we conclude the following equivalence which completes the proof:

$$G_{B_2,B_3} \ \text{ connected with } \ m_2 - 1 \ \text{ h-independent cycles } \ \tilde{\sigma}_1,\dots,\tilde{\sigma}_{m_2-1}$$

$$\Longleftrightarrow \quad G_{rc} \ \text{ connected with } \ m_2 - 1 \ \text{ g-independent cycles } \ \sigma_1,\dots,\sigma_{m_2-1} \ . \quad \square$$

For the design treated in example (3.2.1) the cycles $\sigma_1 = [3, (3,3)_2, 3]$ and $\sigma_2 = [1, (5,1)_2, 5, (5,1)_3, 1]$ turn out to be independent: $g(\sigma_1) = (0 , 1 , 0)^T$, $g(\sigma_2) = (0 , -1 , 1)^T$. By theorem (3.2.2) the design is connected $(m_2 - 1 = 2)$.

3.2.3 Example

Suppose that for $m_1 = 3$, $m_2 = 6$, $m_3 = 11$ a row-column design is given by

$$Y = \begin{bmatrix} 1 & 6 & 9 & 7 & 4 & 10 \\ 3 & 2 & 8 & 5 & 5 & 9 \\ 2 & 8 & 11 & 11 & 3 & 7 \end{bmatrix} .$$

Obviously, this design is connected if and only if the row-column design determined by Y^T is connected (replacing Y by Y^T is nothing but interchanging the roles of factor F_1 and factor F_2) . The connectivity criterion is easier to check if m_2 is small, therefore let G_{rc} be the digraph for Y^T:

G_{rc} :

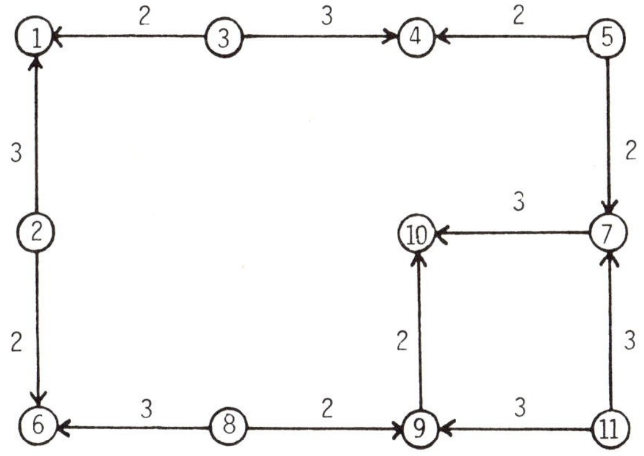

It is easy to see that for each cycle σ in G_{rc} the vector label $g(\sigma)$ is a multiple of $(0, 1, -1)^T$, hence the present row-column design is not connected .

For a row-column design with $m_2 = 2$ all arcs in the digraph G_{rc} have the same label $j = 2$. This shows that the result of Wynn [1977] (cf. theorem (2.3.9)) is a special case of theorem (3.2.2).

Federer/Zelen [1964] conjecture that a row-column design D is connected provided that the block designs given by the matrices D_{12} and D_{13} are connected and $d_1^3 = d_2^3 = \dots = d_{m_3}^3$. The combinatorial nature of theorem (3.2.2) easily allows the construction of counterexamples (cf. also Shah/Khatri [1973]).

3.3 Designs for two-way elimination of heterogeneity

This section deals with general designs D for two-way elimination of heterogeneity (i. e. general three-factor designs) . Therefore the restriction

$$\sum_{k=1}^{m_3} d_{ijk} = 1 \quad \text{for all} \quad 1 \leq i \leq m_1 \quad , \quad 1 \leq j \leq m_2 \quad ,$$

of the preceding section is relaxed. If one replaces this condition by

(3.3.1) $$\sum_{k=1}^{m_3} d_{ijk} > 0 \quad \text{for all} \quad 1 \leq i \leq m_1 \quad , \quad 1 \leq j \leq m_2 \quad ,$$

only a slight modification in the definition of the associated digraph is necessary in order to obtain an analogous result:

Suppose $Y = ((y_{ij}))$ is an $m_1 \times m_2$ "set matrix" , i. e. each entry y_{ij} is a nonempty subset of the set of treatment indices $\{1,\dots,m_3\}$, defined by

$$y_{ij} = \{k \mid 1 \leq k \leq m_3 \, , \, d_{ijk} > 0\}$$

$$1 \leq i \leq m_1 \quad , \quad 1 \leq j \leq m_2 \quad .$$

Now a digraph $\overline{G}$ with integer-valued labels on its arcs is introduced:

$$\overline{G} = (V,\overline{E})$$

$$V = \{1,\dots,m_3\}$$

$$\bar{E} = \{(k,k^*)_o \mid \exists\, i \text{ such that } k^* = m_{i1} < k \in y_{i1}\}$$

$$\cup \{(k,k^*)_j \mid \exists\, i \text{ such that } k^* = m_{i1} , k \in y_{i1} , j \geq 2\}$$

where m_{i1} $(i = 1,\dots,m_1)$ is defined by

$$m_{i1} = \min \{k \mid k \in y_{i1}\} \quad .$$

Vector labels $g(\tau)$ for chains τ in $\bar{G}$ are defined as in the last section:

$$g(\tau) = \sum_{e_j \in \bar{E}(\tau^+)} u^j - \sum_{e_j \in \bar{E}(\tau^-)} u^j$$

As an extension of theorem (3.2.2) we obtain the following

3.3.2 Corollary

A design for two-way elimination of heterogeneity given by a set matrix $Y = ((y_{ij}))$ with the property $\emptyset \neq y_{ij} \subseteq \{1,\dots,m_3\}$ for $i = 1,\dots,m_1$, $j = 1,\dots,m_2$, is connected if and only if the associated digraph $\bar{G}$ is connected and has $m_2 - 1$ independent cycles.

Proof:

The proof is almost identical to that of theorem (3.2.2) , only the structure of the matrix B_2 differing slightly in the case of condition (3.3.1) : additionally rows with all entries zero and repetitions of rows may occur. But the rank of B_2 remains unchanged.

□

For $m_2 = 1$ the second blocking factor becomes redundant; one can look at such a design as a block design (i. e. two-factor design) $\tilde{D} = ((\tilde{d}_{ij}))$ with $\tilde{d}_{ik} = d_{ijk}$ for $i = 1,\dots,m_1$, $k = 1,\dots,m_3$. Orientations and labels of arcs in $\bar{G}$ turn out to be superfluous; it suffices to consider a graph $\tilde{G} = [V,\tilde{E}]$ with

$$\tilde{E} = \{\{k,k^*\} \mid \exists\ i \text{ with } \tilde{d}_{ik} > 0 \text{ and } k^* = \min \{s \mid d_{is} > 0\}\}$$

instead of the digraph $\bar{G} = (V,\bar{E})$. B_2 reduces in the case $m_2 = 1$ to a single column of zeros. It is easy to see that G is a connected graph if and only if the bipartite graph G associated with the block design D (cf. theorem (2.3.2)) is connected. Hence corollary (3.3.2) implies theorem (2.3.2) by considering the special case $m_2 = 1$.

In general the result of theorem (3.2.2) or corollary (3.3.2) cannot be extended to the case where condition (3.3.1) does not hold.

<u>3.3.3 Example</u>

Consider a design D represented by a 4×4 set matrix $Y = ((y_{ij}))$ where $y_{ij} = \{k \mid d_{ijk} > 0\}$:

$$Y = \begin{bmatrix} 1 & 2 & - & - \\ 2 & 1 & - & - \\ - & - & 1 & - \\ - & - & - & 2 \end{bmatrix}$$

(for brevity we omit the parantheses although the entries of Y are sets, and denote the empty set by a dash.) The matrix B of (2.1.1) and P A B of (3.1.1) are

$$B = \begin{bmatrix} 1&0&0&0&1&0&0&0&1&0 \\ 1&0&0&0&0&1&0&0&0&1 \\ 0&1&0&0&1&0&0&0&0&1 \\ 0&1&0&0&0&1&0&0&1&0 \\ 0&0&1&0&0&0&1&0&1&0 \\ 0&0&0&1&0&0&0&1&0&1 \end{bmatrix},$$

$$PAB = \left[\begin{array}{cccc|cccc|cc} 1&0&0&0&1&0&0&0&1&0 \\ 0&1&0&0&1&0&0&0&0&1 \\ 0&0&1&0&0&0&1&0&1&0 \\ 0&0&0&1&0&0&0&1&0&1 \\ \hline 0&0&0&0&1&-1&0&0&1&-1 \\ 0&0&0&0&1&-1&0&0&-1&1 \end{array}\right]$$

$$= \begin{bmatrix} I_4 & \tilde{B}_2 & \tilde{B}_3 \\ 0_{2,4} & B_2 & B_3 \end{bmatrix}.$$

Clearly

$$\pi \in \text{kernel } B = \text{kernel } PAB \implies \pi_1^3 = \pi_2^3$$

therefore by lemma (2.1.9) the design is connected .

$\overline{G}$:

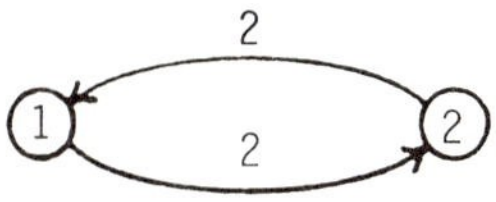

The criterion of corollary (3.2.2) does not apply, we have rank $B_2 = 1$.

As a next step we shall generalize the connectivity characterization to the case of designs for two-way elimination of heterogeneity for which condition (3.3.1) does not necessarily hold.

Consider an arbitrary nonnegative integer $m_1 \times m_2 \times m_3$ matrix D which merely has to fulfil the general requirement of remark (2.1.4) (i. e. there are no "superfluous" factor effects) . Let J denote the set of all indices i $(1 \leq i \leq m_1)$ for which $d_{i1k} = 0$ holds for each treatment index k $(1 \leq k \leq m_3)$, and let $i_1, \dots, i_{|J|}$ be an arrangement of these indices in increasing order, i. e.

$$J = \{i \mid d_{i1}^{12} = 0\}$$

$$= \{i_\nu \mid \nu = 1, \dots, s\}$$

where $i_1 < \dots < i_s$ and $s = |J|$.

The digraph G to be introduced will contain one vertex for each of the $k \in \{1,\dots,m_3\}$ and each $i \in J$. $Y = ((y_{ij}))$ is again taken to be an $m_1 \times m_2$ set matrix with $y_{ij} \subseteq \{1,\dots,m_3\}$, namely the one defined by

$$y_{ij} = \{k \mid d_{ijk} > 0\} \quad , \; i = 1,\dots,m_1 \quad , \; j = 1,\dots,m_2 \quad .$$

Furthermore let

$$m_{ij} = \begin{cases} \min\,\{k \mid k \in y_{ij}\} & \text{if } y_{ij} \neq \emptyset \;; \\ m_3 + \nu & \text{if } y_{ij} = \emptyset \;,\; j = 1 \;,\; i = i_\nu \in J \;; \\ 0 & \text{if } y_{ij} = \emptyset \;,\; j > 1 \;. \end{cases}$$

The following digraph G is associated with D :

$$G = (V,E) \quad ,$$

$$V = \{1,\dots,m_3 + s\} \quad ,$$

$$E = E_1 \cup E_2 \quad ,$$

where

$$E_1 = \{(k,k^*)_0 \mid \exists\; i,j \text{ with } k^* = m_{ij} < k \in y_{ij}\} \quad ,$$

$$E_2 = \{(k,k^*)_j \mid \exists\; i \text{ with } k = m_{ij} > 0 \;,\; k^* = m_{i1} \quad , \; j \geq 2\} \quad .$$

The construction of G can easily be illustrated by an example.

3.3.4 Example

Consider the $3 \times 3 \times 8$ matrix D given by

$$d_{111} = 1 , \quad d_{112} = 2 , \quad d_{113} = 5 , \quad d_{134} = 3 ,$$

$$d_{215} = 1 , \quad d_{226} = 3 , \quad d_{227} = 1 , \quad d_{232} = 2 ,$$

$$d_{323} = 2 , \quad d_{324} = 1 , \quad d_{325} = 7 , \quad d_{338} = 1 ,$$

$$d_{ijk} = 0 \text{ otherwise} .$$

$$Y = \begin{bmatrix} \{1,2,3\} & \emptyset & \{4\} \\ \{5\} & \{6,7\} & \{2\} \\ \emptyset & \{3,4,5\} & \{8\} \end{bmatrix}$$

We have $J = \{i \mid \sum_{k=1}^{m_3} d_{ijk} = 0\} = \{3\} = \{i_1\}$ and for $M = ((m_{ij}))$ we get $m_{31} = m_3 + 1 = 9$, $m_{12} = 0$ and $m_{ij} = \min \{k \mid k \in y_{ij}\}$ otherwise:

$$M = \begin{bmatrix} 1 & 0 & 4 \\ 5 & 6 & 2 \\ 9 & 3 & 8 \end{bmatrix}$$

For each y_{ij} with $|y_{ij}| \geq 2$ the set E contains exactly $|y_{ij}| - 1$ arcs which are labeled 0 , namely

$$(k,m_{ij})_o \quad , \quad k \in y_{ij} \setminus \{m_{ij}\} .$$

The remaining arcs in E are of the form

$$(m_{ij}, m_{il})_j \quad , \quad j \geq 2 \ ,$$

the label j is always identical to the column index of the respective tail vertex.

G :

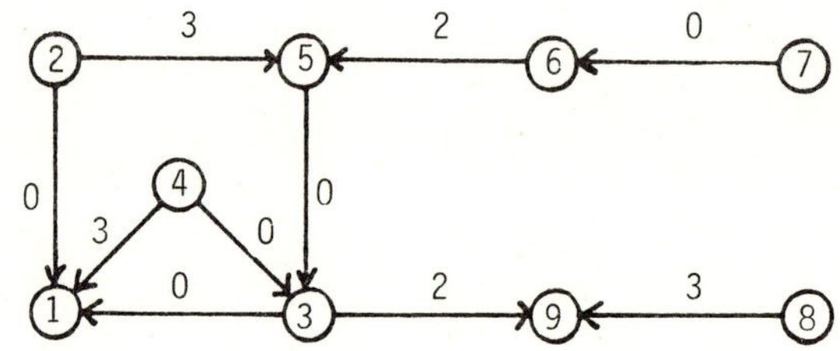

The integer arc labels yield vector labels for the chains τ in G ,

$$\tau \longmapsto g(\tau) \in \mathbb{Z}^{m_2}$$

$$g(\tau) = \sum_{e_j \in E(\tau^+)} u^j - \sum_{e_j \in E(\tau^-)} u^j \ ,$$

where u^j denotes the j-th unit vector of $\mathbb{R}^{m_2}$, $u^o = 0_{m_2}$ (the orientations of the arcs in the set E_1 are therefore irrelevant) .

Another graph is needed for stating the following theorem. Consider the submatrix B_2 in (3.1.1) , i. e. in

$$PAB = \begin{pmatrix} I_{m_1} & \tilde{B}_2 & \tilde{B}_3 \\ 0_{p-m_1,m_1} & B_2 & B_3 \end{pmatrix} .$$

where B is the matrix of the underlying model (2.1.1) for the design D. Now let G_{B_2} be the (undirected) graph associated with the matrix B_2, i. e. a graph of the kind G_Q defined in section 3.1 . It is easy to see how G_{B_2} can be represented by means of the sets y_{ij} :

$$G_{B_2} = [V',E'] ,$$

$$V' = \{1,\dots,m_2\} ,$$

$$E' = \{\{j,j^*\} \mid \exists\, i \text{ with } y_{ij} \neq \emptyset , j > j^* = \min \{\nu \mid y_{i\nu} \neq \emptyset\}\} .$$

For the design in example (3.3.4) the graph G_{B_2} clearly is a complete graph with three vertices. The following connectivity criterion (cf. Butz [1981]) makes use of the number of connected components $c(G_{B_2})$ of this graph.

<u>3.3.5 Theorem</u>

Let a design for two-way elimination of heterogeneity be given by a nonnegative integer $m_1 \times m_2 \times m_3$ matrix D, and consider the associated digraph G and the graph G_{B_2}. D is a connected design if and only if the digraph G is connected and contains $m_2 - c(G_{B_2})$ independent cycles.

Proof:

According to lemma (3.1.3)

$$\text{rank } B_2 = m_2 - c(G_{B_2})$$

hence D is connected if and only if

$$\text{rank } (B_2,B_3) = m_2 - c(G_{B_2}) + m_3 - 1$$

(cf. lemma (3.1.2)) or , by lemma (3.1.6) , if and only if G_{B_2,B_3} is a connected digraph having $m_2 - c(G_{B_2})$ h-independent cycles (cf. the digraph $G_{R,Q}$ of section 3.1 for $R = B_2$, $Q = B_3$) .

We are going to show that for each elementary cycle σ in G there is a cycle $\hat{\sigma}$ in G_{B_2,B_3} , and for each elementary cycle $\hat{\sigma}$ in G_{B_2,B_3} there is a cycle σ in G such that

$$h(\hat{\sigma}) = - g(\sigma) + \lambda \cdot u^1 \quad ,$$

where λ is an appropriate integer. In the digraph G_{B_2,B_3} consider

$$\hat{\tau} = [k',(k',k'')_r,k''] \quad ,$$

i. e. a chain with only one arc $(k',k'')_r$ in G_{B_2,B_3} . Let $r = (u^{j''} - u^{j'})^T$ be the corresponding row of the matrix B_2 , and assume for the moment that $k' \neq k''$. By the construction of (B_2,B_3) and G_{B_2,B_3} we know that

$i \leq j'' \leq j' \leq m_2$ and that there exists an index $i \in \{1,\dots,m_1\}$ such that $k'' = m_{ij''}$ and $m_{ij'} \leq k' \in y_{ij'}$.

Case 1: $1 < j'' < j'$, $m_{ij'} < k'$

$\Longrightarrow (k',m_{ij'})_0$, $(m_{ij'},m_3+\nu)_{j'}$, $(k'',m_3+\nu)_{j''} \in E$ where $i = i_\nu \in J$;

we define $\tau_1 = [k', (k',m_{ij'})_0, m_{ij'}, (m_{ij'},m_3+\nu)_{j'}, m_3+\nu, (k'',m_3+\nu)_{j''}, k'']$ and get $h(\hat{\tau}) = - g(\tau_1)$.

Case 2: $1 < j'' < j'$, $m_{ij'} = k'$

Case 1 applies if one omits the first arc in the chain τ_1.

Case 3: $1 = j'' < j'$, $m_{ij'} < k'$

$\Longrightarrow (k', m_{ij'})_0, (m_{ij'},k'')_{j'} \in E$;

defining $\tau_2 = [k',(k',m_{ij'})_0, m_{ij'}, (m_{ij'}, k'')_{j'}, k'']$ one obtains $h(\hat{\tau}) = - g(\tau_2) + u^1$.

Case 4: $1 = j'' < j'$, $m_{ij'} = k'$

Omitting the first arc in the chain τ_2 yields the same result as in case 3.

Case 5: $j' = j''$

$\Longrightarrow (k',k'')_0 \in E$; here $\tau_3 = [k', (k',k'')_0, k'']$ yields $h(\hat{\tau}) = 0_{m_2} = - g(\tau_3)$.

Now let $k' = k''$. By definition of the digraph G there is $i \in \{1,\dots,m_1\}$ and $k \in \{1,\dots,m_3\}$ such that $k = m_{ij''}$, $m_{ij'} \leq k \in y_{ij'}$. An analoguous examination of cases 1 through 5 yields in each case an elementary cycle in G which has the desired property.

Consequently, an easy induction on the number of arcs in a given elementary cycle $\hat{\sigma}$ in G_{B_2,B_3} shows that one can always find a cycle σ in G such that

$$h(\hat{\sigma}) = - g(\sigma) + \lambda \cdot u^1 \quad (\lambda \in \mathbb{Z}) .$$

To prove the converse, consider first a chain of the form $\tau_1 = [k', (k',k'')_0, k'']$ in G. Let $\hat{E}$ be the arc set of G_{B_2,B_3}.

Case 1: $k'' = m_{ij^*} < k' \in y_{ij}$, $j^* = \min \{j \mid y_{ij} \neq \emptyset\}$

$\Longrightarrow (k',k'')_r \in \hat{E}$ where $r^T = 0_{m_2}$;

defining $\tau_1 = [k', (k',k'')_r, k'']$ yields $h(\hat{\tau}) = - g(\tau_1)$.

Case 2: $k'' = m_{ij^*} < k' \in y_{ij^*}$, $j^* > \tilde{j} = \min \{j \mid y_{ij} \neq \emptyset\}$

$\Longrightarrow (k'',m_{i\tilde{j}})_r , (k',m_{i\tilde{j}})_r \in \hat{E}$ where $r^T = u^{\tilde{j}} - u^{j^*}$;

defining $\tau_2 = [k', (k',m_{i\tilde{j}})_r, m_{i\tilde{j}}, (k'',m_{i\tilde{j}})_r, k'']$ yields

$h(\hat{\tau}_2) = - r^T + r^T = 0_{m_2} = - g(\tau_1)$.

Now consider a chain of the form $\tau_2 = [k', (k',k'')_j, k'']$ in G.

Case 1: $k' \neq k''$

$\Longrightarrow (k',k'')_r \in \hat{E}$ where $r^T = u^1 - u^j$; defining

$\tau = [k', (k',k'')_r, k'']$ yields $h(\hat{\tau}) = - g(\tau_2) + u^1$.

Case 2: $k' = k''$

$\Longrightarrow (k', k')_r \in \hat{E}$ where $r^T = u^1 - u^j$; $\hat{\sigma} = [k', (k',k')_r, k']$ and τ_2 are elementary cycles for which $h(\hat{\sigma}) = - g(\tau_2) + u^1$.

Finally let $\tau_3 = [k', (k',m_3+\nu)_{j'}, m_3+\nu, (k'',m_3+\nu)_{j''}, k'']$ be a chain in G.

Case 1: $j^* = \min \{j \mid y_{i_\nu j} \neq \emptyset\} < j'$, $j^* < j''$

$\Longrightarrow (k',m_{i_\nu j^*})_{r'} , (k'',m_{i_\nu j^*})_{r''} \in \hat{E}$ where $r' = (u^{j^*} - u^{j'})^T$, $r'' = (u^{j^*} - u^{j''})^T$;

defining $\hat{\tau}_1 = [k', (k',m_{i_\nu j^*})_{r'}, m_{i_\nu j^*}, (k'',m_{i_\nu j^*})_{r''}, k'']$ yields $h(\hat{\tau}_1) = - g(\tau_3)$.

Case 2: $j' = \min \{j \mid y_{i_\nu j} \neq \emptyset\}$

$\Longrightarrow (k'',k')_r \in \hat{E}$ where $r^T = u^{j'} - u^{j''}$;

defining $\hat{\tau}_2 = [k', (k'',k')_{r'}, k'']$ yields $h(\hat{\tau}_2) = -g(\tau_3)$.

Case 3: $j'' = \min \{j \mid y_{i_\nu j} \neq \emptyset\}$

$\Longrightarrow (k',k'')_r \in \hat{E}$ where $r^T = u^{j''} - u^{j'}$;

defining $\tau_3 = [k', (k',k'')_r, k'']$ yields $h(\hat{\tau}_3) = -g(\tau_3)$.

Here one also proves by induction on the length of the cycle:
For each elementary cycle σ in G there is a cycle $\hat{\sigma}$ in G_{B_2,B_3} and an integer λ such that $h(\hat{\sigma}) = -g(\sigma) + \lambda \cdot u^1$.

Clearly there cannot be more g-independent cycles than g-independent elementary cycles in G: For each cycle σ in G $g(\sigma)$ is a linear combination of vector labels $g(\sigma_i)$ of elementary cycles σ_i in G . According to section 3.1 the same holds for the cycles in G_{B_2,B_3} and the labeling function h .
Furthermore, $1^T_{m_2} \cdot h(\hat{\sigma}) = 0$ holds for any cycle $\hat{\sigma}$ in G_{B_2,B_3} (also for any chain, cf. the definition of h). Hence altogether the following equivalence holds:

G_{B_2,B_3} is a connected digraph and has exactly ℓ h-independent cycles

$\Longleftrightarrow$ G is a connected digraph and has exactly ℓ g-independent cycles.

This completes the proof of theorem (3.3.5). $\square$

In the following example we consider a general design for two-way elimination of heterogeneity which is given by a 3×3 set matrix $Y = ((y_{ij}))$ (i. e. $y_{ij} = \{k \mid 1 \leq k \leq 8 , d_{ijk} > 0\}$ for $i = 1,2,3$, $j = 1,2,3$) .

3.3.6 Example

$$Y = \begin{bmatrix} 1,2,3 & - & - \\ - & 6,7,8 & 2 \\ - & 3,4,5 & 8 \end{bmatrix}$$

We have $J = \{2,3\}$, $m_{21} = m_3 + 1$, $m_{31} = m_3 + 2$,

$$M = \begin{bmatrix} 1 & 0 & 0 \\ 9 & 6 & 2 \\ 10 & 3 & 8 \end{bmatrix} .$$

The resulting digraph G is connected.

G :

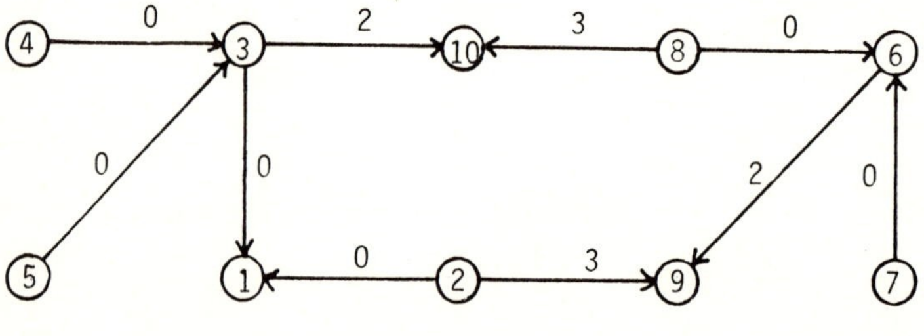

The graph G_{B_2}

G_{B_2} :

has $c(G_{B_2}) = 2$ connected components, and theorem (3.3.5) states that the existence of a single cycle σ in G $(m_2 - c(G_{B_2}) = 1)$ fulfilling $g(\sigma) \neq 0_{m_2}$ is necessary and sufficient for the design to be connected. Such a σ with $g(\sigma) = (0,2,-2)^T$ can easily be read off the digraph.

The connectivity characterization given in theorem (3.3.5) is only one variant of a variety of equivalent criteria which all use digraphs with integer labels on the arcs. An eye-catching modification is obtained by exchanging the roles of the blocking factors F_1 and F_2. Further variants result if one considers digraphs whose vertices do not correspond to the treatments but to the levels of the first or second blocking factor. It turns out that for designs of different dimensionality different characterizations may be preferable, and for this reason we wish to discuss these alternatives in full detail.

Some technical notation is needed. For a given three-factor design D, consider the matrix B of the linear model (2.1.1) . By hypothesis the rows of B are arranged such that the column index triples (i, m_1+j, m_1+m_2+k) of their non-zero entries (exactly three ones per row) are ordered lexicographically (cf. section 2.1). By means of suitable permutations of the rows of B one can also achieve lexicographical orderings of permutations of these column index triples: Let $\bar{P}$ and $\hat{P}$ be $p \times p$ permutation matrices for which the corresponding matrix

$\bar{P}B$ has the property that the index triples (m_1+j, i, m_1+m_2+k) of non-zero row entries are ordered lexicographically

and the matrix

$\hat{P}B$ has the property that the index triples (m_1+m_2+k, i, m_1+j) of non-zero row entries are ordered lexicographically.

Analogously to the definition of the matrices P and A in (3.1.1), i. e. for $n = 3$

$$(3.3.7) \qquad PAB = \begin{bmatrix} I_{m_1} & \tilde{B}_2 & \tilde{B}_3 \\ 0_{p-m_1,m_1} & B_2 & B_3 \end{bmatrix},$$

we define multipliers for $\bar{P}B$ and $\hat{P}B$ which yield appropriate row operations: For $i \in \{2,3\}$ and $j \in \{1,\dots,m_i\}$, denote by A_j^i a $d_j^i \times d_j^i$ matrix of the form

$$A_j^i = \begin{bmatrix} 1 & & & & 0 & \\ 1 & -1 & & & & \\ 1 & & -1 & & & \\ \vdots & & & \ddots & & \\ \vdots & & 0 & & \ddots & \\ 1 & & & & & -1 \end{bmatrix}.$$

All d_j^i are positive by (2.1.4). Using $A_1^2,\dots,A_{m_2}^2$ and $A_1^3,\dots,A_{m_3}^3$ we define $p \times p$ matrices A' and A'' which are both of block diagonal form $(p = \sum_{j=1}^{m_2} d_j^2 = \sum_{k=1}^{m_3} d_k^3)$:

$$A' = \begin{bmatrix} A_1^2 & & & 0 \\ & A_2^2 & & \\ & & \ddots & \\ 0 & & & A_{m_2}^2 \end{bmatrix} ,$$

$$A'' = \begin{bmatrix} A_1^3 & & & 0 \\ & A_2^3 & & \\ & & \ddots & \\ 0 & & & A_{m_3}^3 \end{bmatrix} .$$

Multiplying $\bar{P} B \pi$ from the left with A' , resp. $\hat{P} B \pi$ with A'' , results in an elimination of the parameters π_j^2 , $j = 1,\dots,m_2$, resp. of π_k^3 , $k = 1,\dots,m_3$, by means of suitable row subtractions. Rearranging the rows of $A' \bar{P} B$ and $A'' \hat{P} B$ by multiplying from the left with appropriate $p \times p$ permutation matrices, we can obtain representations which are analogous to (3.1.1) . This is achieved by defining

$$P' = [u^1, u^{s_1'+1}, u^{s_2'+1}, \dots, u^{s_{m_2-1}'+1}, u^2, u^3, \dots, u^{s_1'},$$

$$u^{s_1'+2}, u^{s_1'+3}, \dots, u^{s_{m_2-1}'+2}, \dots, u^{s_{m_2}'}] ,$$

$$P'' = [u^1, u^{s''_1+1}, u^{s''_2+1}, \dots, u^{s''_{m_3-1}+1}, u^2, u^3, \dots, u^{s''_1},$$

$$u^{s''_1+2}, u^{s''_1+3}, \dots, u^{s''_{m_3-1}+2}, \dots, u^{s''_{m_3}}]$$

where

$$s'_\nu = \sum_{j=1}^{\nu} d_j^2 \quad , \quad \nu = 1, \dots, m_2 \quad ,$$

$$s''_\nu = \sum_{k=1}^{\nu} d_k^3 \quad , \quad \nu = 1, \dots, m_3 \quad ,$$

and therefore $p = s'_{m_2} = s''_{m_3}$ (cf. the construction of the matrix P in (3.1.1)). It is easy to verify that this yields the desired representations

(3.3.8) $$P' A' \bar{P} B = \begin{bmatrix} \tilde{B}'_1 & I_{m_2} & \tilde{B}'_3 \\ B'_1 & 0_{p-m_1, m_2} & B'_3 \end{bmatrix}$$

and

(3.3.9) $$P'' A'' \hat{P} B = \begin{bmatrix} \tilde{B}''_1 & \tilde{B}''_2 & I_{m_3} \\ B''_1 & B''_2 & 0_{p-m_3, m_3} \end{bmatrix} .$$

The first blocks of rows $(\tilde{B}'_1, I_{m_3}, \tilde{B}'_2)$ and $(\tilde{B}''_1, \tilde{B}''_2, I_{m_3})$ turn out to be submatrices of B, and each row of the matrices B'_1, B'_3, B''_1 and B''_2 is the difference of two unit vectors of suitable dimension.

For the following lemma we have to define a bipartite graph G_{12} determined by D_{12} (cf. section 2.1 for the definition of D_{ii^*}) :

$$G_{12} = [V_{12}, E_{12}] \quad ,$$

$$V_{12} = V_1 \cup V_2 \quad ,$$

$$V_1 = \{v_1^1, \dots, v_{m_1}^1\} \quad ,$$

$$V_2 = \{v_1^2, \dots, v_{m_2}^2\} \quad ,$$

$$E_{12} = \{\{v_i^1, v_j^2\} \mid d_{ij}^{12} > 0\} \quad .$$

Furthermore let G_{B_2} resp. $G_{B_1'}$ be the graph G_Q of section 3.1 for $Q = B_2$ and $Q = B_1'$, respectively.

3.3.10 Lemma

$c(G_{12}) = c(G_{B_2}) = c(G_{B_1'})$, i. e. the graphs G_{12} , G_{B_2} and $G_{B_1'}$ have the same number of connected components.

Proof:

Define an $m_1 \times m_2$ matrix $N = ((n_{ij}))$ by

$$n_{ij} = \begin{cases} 1 & \text{if } d_{ij}^{12} > 0 \quad , \\ 0 & \text{otherwise} \quad , \end{cases}$$

$$i = 1, \dots, m_1 \quad , \quad j = 1, \dots, m_2 \quad .$$

N turns out to be the adjacency matrix of the bipartite graph G_{12} , i. e. $v_i^1 \in V_1$ is adjacent to $v_j^2 \in V_2$ if and only if $n_{ij} = 1$. Let $G_{B_1'} = [V_1, E_{B_1'}]$ and $G_{B_2} = [V_2, E_{B_2}]$. For $1 \leq i' < i'' \leq m_1$ and $1 \leq j' < j'' \leq m_2$ we know by construction of B_1' and B_2:

$$\{v^1_{i'}, v^1_{i''}\} \in E_{B'_1} \iff \exists\, j^* \quad n_{i'j^*} = n_{i''j^*} = 1 \text{ and}$$

$$n_{ij^*} = 0 \text{ for all } 1 \leq i < i'$$

and

$$\{v^2_{j'}, v^2_{j''}\} \in E_{B_2} \iff \exists\, i^* \quad n_{i^*j'} = n_{i^*j''} = 1 \text{ and}$$

$$n_{i^*j} = 0 \text{ for all } 1 \leq j < j' \quad .$$

This implies: If $v^1_{i'}$ and $v^1_{i''}$ are linked by a chain in $G_{B'_1}$, then they are also linked by a chain in G_{12}. If $v^2_{j'}$ and $v^2_{j''}$ are linked by a chain in G_{B_2}, then the same holds in G_{12}. It follows that

$$c(G_{B'_1}) \geqq c(G_{12})$$

and

$$c(G_{B_2}) \geqq c(G_{12}) \quad .$$

Suppose $c(G_{B'_1}) > c(G_{12})$. Then V_1 contains two vertices which are linked by a chain in G_{12}, but not in $G_{B'_1}$. On this chain in G_{12} there are vertices $v^1_{i_1}, v^1_{i_2} \in V_1$ which belong to different connected components of $G_{B'_1}$ and which are both adjacent to some $v^2_{j^*} \in V_2$ in G_{12}, i. e. $n_{i_1j^*} = n_{i_2j^*} = 1$. This leads to a contradiction: take $\min\{i \mid n_{ij^*} = 1\} = i^*$, then the edges $\{v^1_{i^*}, v^1_{i_1}\}$ and $\{v^1_{i^*}, v^1_{i_2}\}$ must belong to $E_{B'_1}$, i. e. $v^1_{i_1}$ and $v^1_{i_2}$ are in the same connected component of $G_{B'_1}$.

Suppose $c(G_{B_2}) > c(G_{12})$. Here one arrives at a contradiction analogously, as every pair of vertices $v^2_{j_1}, v^2_{j_2} \in V_2$ with $\{v^1_{i^*}, v^2_{j_1}\}, \{v^1_{i^*}, v^2_{j_2}\} \in E_{12}$

is linked by a chain in G_{B_2} : take $\min\{j \mid n_{i^*j} = 1\} = j^*$, then it is clear that $\{v^2_{j^*} , v^2_{j_1}\}$, $\{v^2_{j^*} , v^2_{j_2}\} \in E_{B_2}$. $\square$

The digraph G defined for theorem (3.3.5) is closely related to the digraph $G_{R,Q}$ of lemma (3.1.6) where $R = B_2$, $Q = B_3$, cf. the proof of theorem (3.3.5) . For a first obvious variant of the characterization in theorem (3.3.5) we derive a digraph from G_{B_3,B_2} , i. e. from $G_{R,Q}$ where $R = B_3$, $Q = B_2$, the submatrices B_2 and B_3 in (3.3.7) having switched their roles. The resulting digraph is called

$$G^2_3 = (V^2_3, E^2_3) \quad .$$

The upper index 2 reflects the fact that the vertex set in G_{B_3,B_2} is given by the indices $1,\dots,m_2$ of the levels of factor F_2 . The lower index 3 points out that the arc labels correspond to the treatments, i. e. to the levels of the factor F_3 . For the sake of consistency the digraph G of theorem (3.3.5) will from now on be denoted by

$$G = G^3_2 = (V^3_2, E^3_2) \quad .$$

In the sequel of this section we will also examine connectivity characterizations due to the digraphs $G_{B'_1,B'_3}$, $G_{B'_3,B'_1}$, $G_{B''_1,B''_2}$ and $G_{B''_2,B''_1}$ (B'_1 and B'_3 are submatrices of $P'\,A'\,\bar{P}\,B$ in (3.3.8), B''_1 and B''_2 are submatrices of $P''\,B''\,\hat{P}\,B$ in (3.3.9)). The corresponding digraphs actually used in the connectivity criteria to be presented are G^3_1 , G^1_3 , G^2_1 and G^1_2 , therefore we first give a general definition for G^s_t ($s \neq t$, $s,t \in \{1,2,3\}$) which also subsumes the digraphs G_2^3 and G_3^2 :

For fixed distinct $s,t \in \{1,2,3\}$ let r be the remaining element in $\{1,2,3\} \setminus \{s,t\}$. Now let

$$J^{rt} = \begin{cases} \{i \mid d_{i1}^{rt} = 0\} & \text{if } r < t \\ \{i \mid d_{1i}^{tr} = 0\} & \text{if } t < r \end{cases}$$

and let $i_1,\dots,i_{|J^{rt}|}$ be an arrangement of these indices in increasing order, i. e.

$$J^{rt} = \{i_\nu \mid \nu = 1,\dots,|J^{rt}|\}$$

$$\text{where } i_1 < \dots < i_{|J^{rt}|} .$$

Furthermore define an $m_r \times m_t$ set matrix $Y^{rt} = ((y_{ij}^{rt}))$ by

$$y_{ij}^{rt} = \{k \mid d_{\ell_1\ell_2\ell_3} > 0 \text{ where } i = \ell_r \text{ , } j = \ell_t \text{ and } k = \ell_t\}$$

$$i = 1,\dots,m_r \quad , \quad j = 1,\dots,m_t \quad .$$

Additionally we need

$$m_{ij}^{rt} = \begin{cases} \min\{k \mid k \in y_{ij}^{rt}\} & \text{if } y_{ij}^{rt} \neq \emptyset \text{ ,} \\ m_s + \nu & \text{if } y_{ij}^{rt} = \emptyset \text{ , } j = 1 \text{ and } i = i_\nu \in J^{rt} \text{ ,} \\ 0 & \text{if } y_{ij}^{rt} = \emptyset \text{ and } j > 1 \text{ ,} \end{cases}$$

where $i = 1,\dots,m_r$, $j = 1,\dots,m_t$. The $m_r \times m_t$ matrix $((m_{ij}^{rt}))$ is used for defining the digraph G_t^s :

$$G_t^s = (V_t^s, E_t^s)$$

$$V_t^s = \{1,\dots,m_s + |J^{rt}|\}$$

$$E_t^s = \{(k,k^*)_o \mid \exists\ i,j \text{ with } k^* = m_{ij}^{rt} < k \in y_{ij}^{rt}\}$$

$$\cup \{(k,k^*)_j \mid \exists\ i \text{ with } k = m_{ij}^{rt} > 0\ ,\ k^* = m_{i1}^{rt}\ ,\ j \geqq 2\}$$

It is easy to see that G_2^3 is the digraph appearing in theorem (3.3.5) , because $t = 2$ and $s = 3$ imply $J^{12} = J$, $m_{ij}^{12} = m_{ij}$ and $y_{ij}^{12} = y_{ij}$ $(i = 1,\dots,m_1\ ,\ j = 1,\dots,m_2)$.

For each y_{ij}^{rt} with $|y_{ij}| \geqq 2$ the set E_t^s contains exactly $|y_{ij}^{rt}| - 1$ arcs which are labeled 0 , namely

$$(k,m_{ij}^{rt})_o\ ,\ k \in y_{ij}^{rt} \setminus \{m_{ij}^{rt}\}\ .$$

The remaining arcs in E_t^s are of the form

$$(m_{ij}^{rt}, m_{i1}^{rt})_j\ ,\ j \geqq 2\ ,$$

and their labels are identical to the column indices j of the respective tail vertices. For chains τ in G_t^s we define vector labels $g(\tau) \in \mathbb{R}^{m_s}$ by means of the integer arc labels:

$$g(\tau) = \sum_{e_j \in E_t^s(\tau^+)} u^j - \sum_{e_j \in E_t^s(\tau^-)} u^j$$

where $u^o = 0_{m_s}$ and u^j is the j-th unit vector of $\mathbb{R}^{m_s}$.

For the following theorem we consider the case $s = 2$, $t = 3$.

3.3.11 Theorem

A design D for two-way elimination of heterogeneity is connected if and only if the associated digraph G_3^2 contains $m_3 - 1$ independent cycles.

Proof:

By lemma (3.1.2) the design is connected if and only if

$$\text{rank } (B_2, B_3) = m_3 - 1 + \text{rank } B_2 \quad .$$

This condition is satisfied if and only if there exist exactly $m_3 - 1$ h-independent cycles in the digraph G_{B_3,B_2} (cf. lemma (3.1.5) , taking $R = B_3$ and $Q = B_2$) .

In perfect analogy to the proof of theorem (3.3.5) one can verify that the maximum number of independent cycles in G_{B_3,B_2} equals the maximum number of independent cycles in G_3^2 .
□

In theorem (3.3.5) the connectedness of the digraph G (= G_2^3) is a necessary connectivity condition for the design D . This cannot be applied to the digraph G_3^2 : we have

$$c(G_3^2) = c(G_{B_2}) = c(G_{12})$$

whether the design D is connected or not (cf. lemma (3.3.10)) .

In the next result we have $t = 1$, $s = 3$.

3.3.12 Corollary

For a design D for two-way elimination of heterogeneity consider the digraph G_1^3 and the bipartite graph G_{12} . D is a connected design if and only if G_1^3 is connected and contains $m_1 - c(G_{12})$ independent cycles.

Proof:

It is easy to see that G_1^3 can be obtained from G_2^3 merely by exchanging the roles of the blocking factors F_1 and F_2 in the design D . Therefore lemma (3.3.10) implies the assertion of the corollary because the design remains connected under the above exchange. A direct proof makes use of lemma (3.1.2) and (3.1.6) : take $R = B_1'$, $Q = B_3'$ and consider the digraph $G_{B_1',B_3'}$. □

Exchanging the roles of F_1 and F_2 in theorem (3.3.11) yields an analogous corollary with $s = 1$, $t = 3$.

3.3.13 Corollary

A design D for two-way elimination of heterogeneity is connected if and only if the digraph G_3^1 contains $m_3 - 1$ independent cycles.

Proof:

Follows from an easy symmetry argument. A direct proof makes use of the digraph $G_{B_3', B_1'}$ (cf. lemma (3.1.5) with $R = B_3'$, $Q = B_1'$) . □

Two further connectivity criteria can be derived by means of the submatrices B_1'' and B_2'' of $P'' A'' \hat{P} B$ in (3.3.9) . We first mention a necessary condition which uses the graphs $G_{B_1''}$ and $G_{B_2''}$ (cf. lemma (3.1.3) for $Q = B_1''$ and $Q = B_2''$ respectively).

3.3.14 Lemma

Consider a design D for two-way elimination of heterogeneity and the graphs $G_{B_1''}$ and $G_{B_2''}$ determined by D . If D is a connected design, then $G_{B_1''}$ and $G_{B_2''}$ are connected graphs.

Proof:

Suppose $G_{B_1''}$ is not connected. We define an $m_1 \times m_3$ matrix $N = ((n_{ik}))$ by

$$n_{ik} = \begin{cases} 1 & \text{if } d_{ik}^{13} > 0 , \\ 0 & \text{otherwise} , \end{cases}$$

$$i = 1, \ldots, m_1 \quad , \quad k = 1, \ldots, m_3 .$$

For $G_{B_1''} = [V_1, E_1]$ we have $V_1 = \{1, \ldots, m_1\}$ and (by construction of B_1'')

$$E_1 = \{\{i',i''\} \mid i' \neq i'' , \exists k \ \ n_{i'k} n_{i''k} = 1 ,$$
$$\forall i < \min\{i',i''\} \ \ n_{ik} = 0 \} .$$

Without loss of generality we assume that N can be represented in the following way :

(3.3.15) $$N = \left(\begin{array}{c|c} N_1 & 0 \\ \hline 0 & N_2 \end{array} \right)$$

and that $n_{11} = n_{m_1 m_3} = 1$ (this can always be achieved by permuting the row and column indices, as $G_{B_2''}$ is not connected) . Here N_1 is an $\bar{m}_1 \times \bar{m}_3$ submatrix of N , $1 \leq \bar{m}_1 < m_1$, $1 \leq \bar{m}_3 < m_3$. Now consider the following parameter vector $\pi = ((\pi^1)^T, (\pi^2)^T, (\pi^3)^T)^T$:

$$\pi_i^1 = \begin{cases} 1 & \text{for} \quad i = 1,\dots,\bar{m}_1 , \\ 0 & \text{for} \quad i = \bar{m}_1 + 1,\dots,m_1 , \end{cases}$$

$$\pi_j^2 = \quad 0 \quad \text{for} \quad j = 1,\dots,m_2 ,$$

$$\pi_k^3 = \begin{cases} -1 & \text{for} \quad k = 1,\dots,\bar{m}_3 , \\ 0 & \text{for} \quad k = \bar{m}_3 + 1,\dots,m_3 . \end{cases}$$

It follows readily that

$$\pi_i^1 + \pi_k^3 = 0 \quad \forall \, i,k \ \text{ with } \ n_{ik} = 1$$

and even

$$\pi_i^1 + \pi_j^2 + \pi_k^3 = 0 \qquad \forall\ i,j,k \text{ with } d_{ijk} > 0 \quad ,$$

the latter being equivalent to $B\,\pi = 0$. Hence by lemma (2.1.9) the design cannot be connected $(\pi_1^3 \neq \pi_{m_3}^3)$.

If $G_{B_2''}$ is not a connected graph, define an $m_2 \times m_3$ matrix $N = ((n_{jk}))$:

$$n_{jk} = \begin{cases} 1 & \text{if } d_{jk}^{23} > 0 \quad , \\ 0 & \text{otherwise} \quad , \end{cases}$$

$$j = 1,\dots,m_2 \quad , \quad k = 1,\dots,m_3 \quad .$$

The vertex set of $G_{B_2''} = [V_2,E_2]$ is $V_2 = \{1,\dots,m_2\}$ and the edge set (by construction of B_2'')

$$E_2 = \Big\{\{j',j''\} \mid j' \neq j'' \ , \ \exists\ k \ \ n_{j'k}\, n_{j''k} = 1 \ ,$$
$$\forall\ j < \min\{j',j''\} \ \ n_{jk} = 0 \Big\} \ .$$

Again we can assume that N has a block representation like (3.3.15) , where now N_1 is an $\bar{m}_2 \times \bar{m}_3$ submatrix, $1 \leq \bar{m}_2 < m_2$, $1 \leq \bar{m}_3 < m_3$. Using the same argument as in the first part of the proof, we can specify an element of the kernel of B which again leads to a contradiction:

$$\pi_i^1 = 0 \qquad \text{for } i = 1,\dots,m_1 \quad ,$$

$$\pi_j^2 = \begin{cases} 1 & \text{for } i = 1,\dots,\bar{m}_2 \quad , \\ 0 & \text{for } i = \bar{m}_2 + 1,\dots,m_2 \quad , \end{cases}$$

$$\pi_k^3 = \begin{cases} -1 & \text{for } k = 1,\dots,\bar{m}_3 \ , \\ 0 & \text{for } k = \bar{m}_3+1,\dots,m_3 \ . \end{cases}$$ □

The next necessary and sufficient criterion makes use of G_{12} and G_t^s where $t = 1$ and $s = 2$.

3.3.16 Theorem

For a design D for two-way elimination of heterogeneity consider the bipartite graph G_{12} and the digraph G_1^2 . D is a connected design if and only if G_1^2 is connected and contains $m_1 - c(G_{12})$ independent cycles.

Proof:

By (3.3.7) and lemmas (3.1.2) and (3.3.10) we know that D is connected if and only if

$$\text{rank } B = m_1 + m_3 - 1 + m_2 - c(G_{12}) \ .$$

From (3.3.9) it follows that

$$\text{rank } B = \text{rank } P''A''\hat{P}\,B = m_3 + \text{rank } (B_1'',B_2'') \ ,$$

therefore

$$\text{rank } (B_1'',B_2'') = m_2 - 1 + m_1 - c(G_{12})$$

is a necessary and sufficient condition for the connectivity of D . Applying lemma (3.1.5) to the digraph $G_{B''_1,B''_2}$ (i. e. $R = B''_1$, $Q = B''_2$) and keeping in mind that $c(G_{B''_1,B''_2}) = c(G_{B''_2})$, lemmas (3.1.3) and (3.3.14) yield

D is a connected design

$\Longrightarrow$ rank $B''_2 = m_2 - 1$, the digraph $G_{B''_1,B''_2}$ is connected and contains exactly $m_1 - c(G_{12})$ h-independent cycles.

Conversely we have the implication

$G_{B''_1,B''_2}$ is a connected digraph with $m_1 - c(G_{12})$ h-independent cycles

$\Longrightarrow$ rank $B''_2 = m_2 - 1$, rank $(B''_1,B''_2) = m_2 - 1 + m_1 - c(G_{12})$.

In analogy to the proof of theorem (3.3.5) one can verify that G_1^2 is a connected digraph if and only if $G_{B''_1,B''_2}$ is a connected digraph, and that G_1^2 contains exactly $m_1 - c(G_{12})$ g-independent cycles if and only if $G_{B''_1,B''_2}$ has $m_1 - c(G_{12})$ h-independent cycles. $\square$

Finally, the last variant of G_t^s to be examined is G_2^1 .

3.3.17 Corollary

Consider the digraph G_2^1 and the bipartite graph G_{12} determined by a design D for two-way elimination of heterogeneity. D is a connected design if and only if G_2^1 is a connected digraph which contains $m_2 - c(G_{12})$ independent cycles.

Proof:

The digraph G_2^1 can be obtained from G_1^2 merely by exchanging the roles of the blocking factors F_1 and F_2 in the design D . Therefore the corollary follows immediately from theorem (3.3.16) . A direct proof makes use of the digraph $G_{B_2'', B_1''}$, cf. lemma (3.1.5) with $R = B_2''$ and $Q = B_1''$. □

The connectivity characterizations of the theorems (3.3.5) , (3.3.11) , (3.3.16) and the corollaries (3.3.12) , (3.3.13) , (3.3.17) will now be illustrated by the example which has already been treated in (3.3.6) .

(3.3.18) Example

Consider a design D for two-way elimination of heterogeneity , where $m_1 = m_2 = 3$, $m_3 = 8$ and

$$d_{111} = 2 \ , \quad d_{112} = 3 \ , \quad d_{113} = 1 \ ,$$

$$d_{226} = 5 \ , \quad d_{227} = 1 \ , \quad d_{228} = 1 \ ,$$

$$d_{232} = 1 \ , \quad d_{323} = 6 \ , \quad d_{324} = 2 \ ,$$

$$d_{325} = 3 \ , \quad d_{338} = 1 \ , \quad d_{ijk} = 0 \text{ otherwise } .$$

Applying the definitions of J^{rt} , y_{ij}^{rt} , m_{ij}^{rt} and G_t^s (cf. pp 80/81) we get the following six cases :

(i) $r = 1 , t = 2 , s = 3$

$$J^{12} = \{2,3\} \quad , \quad ((y_{ij}^{12})) = \begin{bmatrix} 1,2,3 & - & - \\ - & 6,7,8 & 2 \\ - & 3,4,5 & 8 \end{bmatrix}$$

$$((m_{ij})) = \begin{bmatrix} 1 & 0 & 0 \\ 9 & 6 & 2 \\ 10 & 3 & 8 \end{bmatrix}$$

G_2^3 :

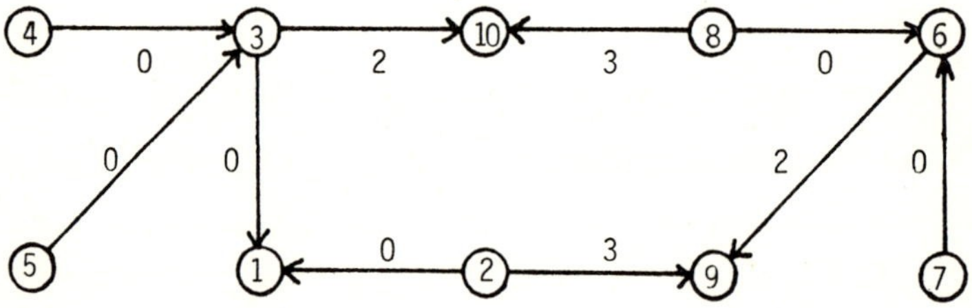

(ii) $r = 1 , s = 2 , t = 3$

$$J^{13} = \{2,3\} \quad , \quad ((y_{ij}^{13})) = \begin{bmatrix} 1 & 1 & 1 & - & - & - & - & - \\ - & 3 & - & - & - & 2 & 2 & 2 \\ - & - & 2 & 2 & 2 & - & - & 3 \end{bmatrix}$$

$$((m_{ij}^{13})) = \begin{bmatrix} 1 & 1 & 1 & 0 & 0 & 0 & 0 & 0 \\ 4 & 3 & 0 & 0 & 0 & 2 & 2 & 2 \\ 5 & 0 & 2 & 2 & 2 & 0 & 0 & 3 \end{bmatrix}$$

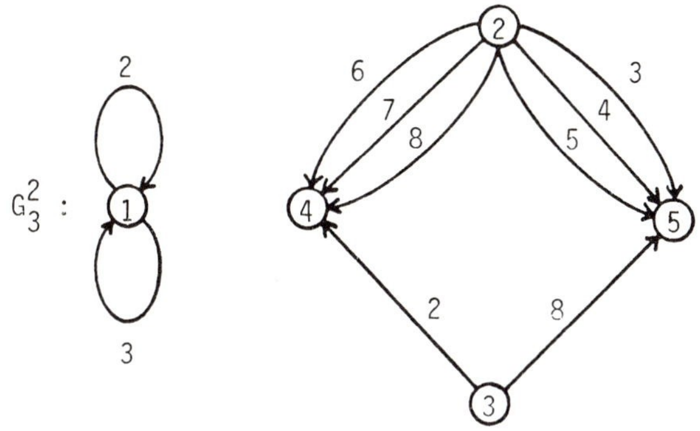

(iii) $t = 1$, $r = 2$, $s = 3$

$$J^{21} = \{2,3\} \ , \ ((y_{ij}^{21})) = ((y_{ij}^{12}))^T \ , \ ((m_{ij}^{21})) = \begin{bmatrix} 1 & 0 & 0 \\ 9 & 6 & 3 \\ 10 & 2 & 8 \end{bmatrix}$$

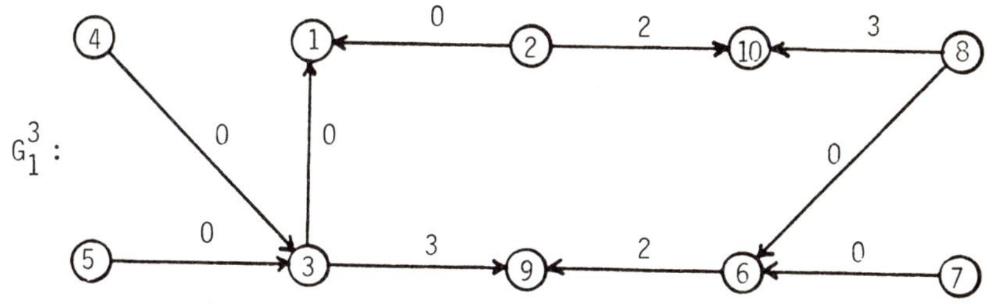

(iv) $s = 1$, $r = 2$, $t = 3$

$$J^{23} = \{2,3\} \ , \ ((y_{ij}^{23})) = \begin{bmatrix} 1 & 1 & 1 & - & - & - & - & - \\ - & - & 3 & 3 & 3 & 2 & 2 & 2 \\ - & 2 & - & - & - & - & - & 3 \end{bmatrix}$$

$$((m_{ij}^{23})) = \begin{bmatrix} 1 & 1 & 1 & 0 & 0 & 0 & 0 & 0 \\ 4 & 0 & 3 & 3 & 3 & 2 & 2 & 2 \\ 5 & 2 & 0 & 0 & 0 & 0 & 0 & 3 \end{bmatrix}$$

G_3^1 :

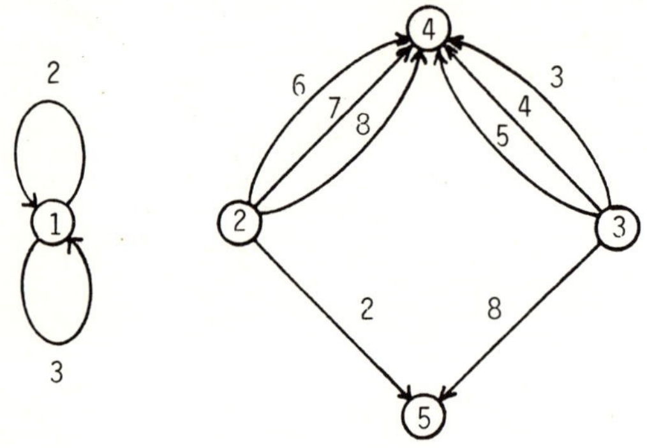

(v) t = 1 , s = 2 , r = 3

$$J^{31} = \{4,5,6,7,8\} \quad , \quad ((y_{ij}^{31})) = ((y_{ij}^{13}))^T \; , \; ((m_{ij}^{31})) = \begin{bmatrix} 1 & 0 & 0 \\ 1 & 3 & 0 \\ 1 & 0 & 2 \\ 4 & 0 & 2 \\ 5 & 0 & 2 \\ 6 & 2 & 0 \\ 7 & 2 & 0 \\ 8 & 2 & 3 \end{bmatrix}$$

G_1^2 :

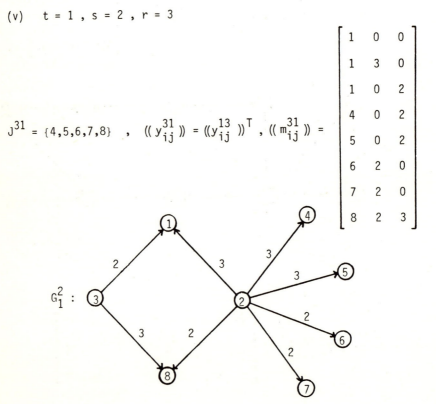

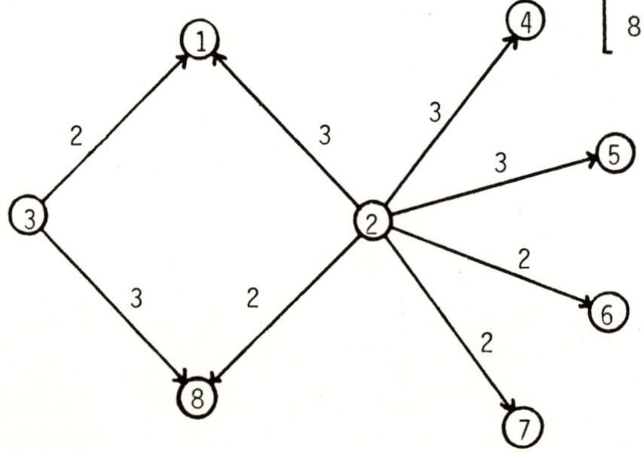

(vi) $s = 1$, $t = 2$, $r = 3$

$$J^{32} = \{4,5,6,7,8\} \quad , \quad ((y^{32}_{ij})) = ((y^{23}_{ij}))^T , \quad ((m^{32}_{ij})) = \begin{bmatrix} 1 & 0 & 0 \\ 1 & 0 & 2 \\ 1 & 3 & 0 \\ 4 & 3 & 0 \\ 5 & 3 & 0 \\ 6 & 2 & 0 \\ 7 & 2 & 0 \\ 8 & 2 & 3 \end{bmatrix}$$

G^1_2 :

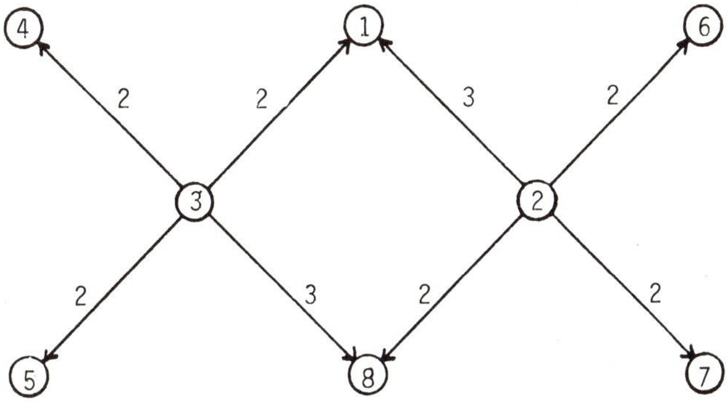

Finally, the bipartite graph G_{12} determined by D is the following (cf. lemma (3.3.10)):

G_{12} :

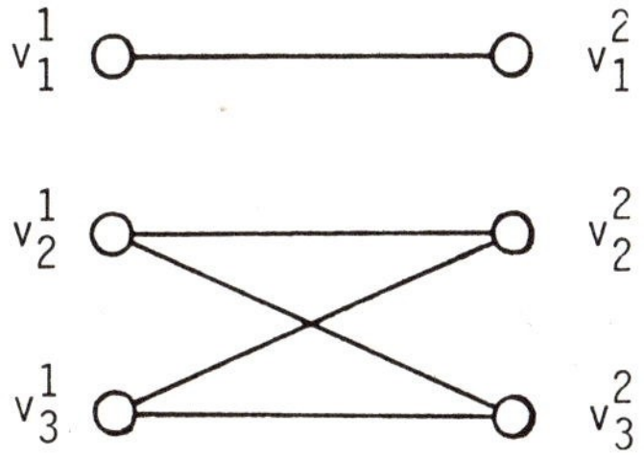

We have $m_1 - c(G_{12}) = m_2 - c(G_{12}) = 1$, hence

D is a connected design

$\Longleftrightarrow$ G_2^3 is a connected digraph containing a cycle which has a nonzero vector label (theorem (3.3.5)) ,

$\Longleftrightarrow$ the digraph G_3^2 has $m_3 - 1 = 7$ independent cycles (theorem (3.3.11)) ,

$\Longleftrightarrow$ G_1^3 is a connected digraph containing a cycle with a nonzero vector label (corollary (3.3.12)) ,

$\Longleftrightarrow$ the digraph G_3^1 has $m_3 - 1 = 7$ independent cycles (corollary (3.3.13)) ,

$\Longleftrightarrow$ G_1^2 has a connected digraph containing a cycle with a nonzero vector label (theorem (3.3.16)) ,

$\Longleftrightarrow$ G_2^1 is a connected digraph containing a cycle with a nonzero vector label (corollary (3.3.17)) .

In the above example it is more tedious to check the connectivity criteria by means of G_3^1 and G_3^2 than by means of the other digraphs. In general it is preferable to work with a criterion demanding the existence of only a small number of independent cycles, furthermore for simplicity one should prefer digraphs having only a small number of vertices. This suggest the following choice (take $c = c(G_{12})$) :

- G_2^3 (theorem (3.3.5)) , if $m_2 - c \leq \min\{m_1 - c, m_3 - 1\}$ and $m_3 \leq m_1$;
- G_2^1 (corollary (3.3.17)) , if $m_2 - c \leq \min\{m_1 - c, m_3 - 1\}$ and $m_1 \leq m_3$;
- G_3^2 (theorem (3.3.11)) , if $m_3 - 1 \leq \min\{m_1 - c, m_2 - c\}$ and $m_2 \leq m_1$;
- G_3^1 (corollary (3.3.13)) , if $m_3 - 1 \leq \min\{m_1 - c, m_2 - c\}$ and $m_1 \leq m_2$;
- G_1^3 (corollary (3.3.12)) , if $m_1 - c \leq \min\{m_2 - c, m_3 - 1\}$ and $m_3 \leq m_2$;
- G_1^2 (theorem (3.3.16)) , if $m_1 - c \leq \min\{m_2 - c, m_3 - 1\}$ and $m_2 \leq m_3$.

In chapter 4 invariance properties of these digraphs will be introduced which simplify the investigation of the corresponding criteria.

By definition (2.1.13) the notation of *connectivity* is synonymous with F_3-*connectivity* for designs for two-way elimination of heterogeneity. In what follows the relations between F_1- , F_2- and F_3-connectivity will be studied. We also examine the block designs D_{12} , D_{23} and D_{13} *induced* by a given design for two-way elimination of heterogeneity (cf. section 2.1 for the definition if D_{ii^*} , $i < i^*$) .

In analogy to the definition of the bipartite graph G_{12} (cf. lemma (3.3.10)) , whose adjacency structure is determined by the matrix D_{12} , we define bipartite graphs which are associated with the matrices D_{23} and D_{13} . Let

$$V_i = \{v_1^i,\dots,v_{m_i}^i\}$$

denote sets of vertices (i = 1,2,3) , then a bipartite graph defined by D_{23} is

$$G_{23} = [V_{23},E_{23}]$$

where

$$V_{23} = V_2 \cup V_3 ,$$

$$E_{23} = \{\{v_j^2,v_k^3\} \mid d_{jk}^{23} > 0\} .$$

Analogously also D_{13} defines a bipartite graph, namely

$$G_{13} = [V_{13},E_{13}]$$

where

$$V_{13} = V_1 \cup V_3 \quad ,$$

$$E_{13} = \{\{v_i^1, v_k^3\} \mid d_{ik}^{13} > 0\} \quad .$$

<u>3.3.19 Theorem</u>

Consider the block designs D_{12} , D_{23} and D_{13} induced by a given three-factor design D and the associated bipartite graphs G_{12} , G_{23} and G_{13} . Furthermore consider the graphs G_{B_2} , G_{B_3} , $G_{B_1'}$ $G_{B_3'}$, $G_{B_1''}$ and $G_{B_2''}$ which are determined by submatrices taken from (3.3.7) through (3.3.9) .

(a) Conditions (i) - (vi) are equivalent:

(i) D_{12} is a connected block design ,

(ii) G_{12} is a connected graph ,

(iii) G_{B_2} is a connected graph ,

(iv) $G_{B_1'}$ is a connected graph ,

(v) G_3^2 is a connected digraph ,

(vi) G_3^1 is a connected digraph .

(b) Conditions (vii) - (xii) are equivalent:

(vii) D_{23} is a connected block design ,

(viii) G_{23} is a connected graph ,

(ix) $G_{B'_3}$ is a connected graph ,

(x) $G_{B''_2}$ is a connected graph ,

(xi) G_1^3 is a connected digraph ,

(xii) G_1^2 is a connected digraph .

(c) Conditions (xiii) - (xviii) are equivalent:

(xiii) D_{13} is a connected block design ,

(xiv) G_{13} is a connected graph ,

(xv) G_{B_3} is a connected graph ,

(xvi) $G_{B''_1}$ is a connected graph ,

(xvii) G_2^3 is a connected digraph ,

(xviii) G_2^1 is a connected digraph .

Proof:

Theorem (2.3.2) implies (i) $\Longleftrightarrow$ (ii) , (vii) $\Longleftrightarrow$ (viii) and (xiii) $\Longleftrightarrow$ (xiv) . Note that

$$c(G_{R,Q}) = c(G_Q) \quad ,$$

hence (iii) $\Longleftrightarrow$ (v) and (iv) $\Longleftrightarrow$ (vi) are implied by the proofs of theorem (3.3.11) and corollary (3.3.13) respectively, (ix) $\Longleftrightarrow$ (xi) and (x) $\Longleftrightarrow$ (xii) by the proofs of corollary (3.3.12) and theorem (3.3.16) respectively, and (xv) $\Longleftrightarrow$ (xvii) and (xvi) $\Longleftrightarrow$ (xviii) by the proofs of theorem (3.3.5) and corollary (3.3.17) respectively. Lemma (3.3.10) implies (ii) $\Longleftrightarrow$ (iii) $\Longleftrightarrow$ (iv) . In analogy to lemma (3.3.10) one can also prove that

(3.3.20) $$c(G_{23}) = c(G_{B_3'}) = c(G_{B_2''}) \quad ,$$

(3.3.21) $$c(G_{13}) = c(G_{B_3}) = c(G_{B_1''}) \quad .$$

Now (3.3.20) implies (viii) $\Longleftrightarrow$ (ix) $\Longleftrightarrow$ (x) and finally (xiv) $\Longleftrightarrow$ (xv) $\Longleftrightarrow$ (xvi) follows from (3.3.21) . $\square$

In table (3.3.22) the notation

$$z(G_t^s) \qquad (s \neq t \in \{1,2,3\})$$

denotes the maximum number of independent cycles in the digraph G_t^s . The last column repeats the results of theorems (3.3.5) , (3.3.11) , (3.3.16) and corollaries (3.3.12) , (3.3.13) , (3.3.17) . The digraphs G_t^s can also be used to derive necessary and sufficient conditions for F_1- and F_2-connectivity; the resulting criteria can be found in the first and second column of table (3.3.22) .

	F_1-connectivity	F_2-connectivity	F_3-connectivity
G_2^1	$z(G_2^1) = m_2 - c(G_{23})$ $c(G_2^1) = 1$	$z(G_2^1) = m_2 - 1$	$z(G_2^1) = m_2 - c(G_{12})$ $c(G_2^1) = 1$
G_1^2	$z(G_1^2) = m_1 - 1$	$z(G_1^2) = m_1 - c(G_{13})$ $c(G_1^2) = 1$	$z(G_1^2) = m_1 - c(G_{12})$ $c(G_1^2) = 1$
G_3^1	$z(G_3^1) = m_3 - c(G_{23})$ $c(G_3^1) = 1$	$z(G_3^1) = m_3 - c(G_{13})$ $c(G_3^1) = 1$	$z(G_3^1) = m_3 - 1$
G_1^3	$z(G_1^3) = m_1 - 1$	$z(G_1^3) = m_1 - c(G_{13})$ $c(G_1^3) = 1$	$z(G_1^3) = m_1 - c(G_{12})$ $c(G_1^3) = 1$
G_3^2	$z(G_3^2) = m_3 - c(G_{23})$ $c(G_3^2) = 1$	$z(G_3^2) = m_3 - c(G_{13})$ $c(G_3^2) = 1$	$z(G_3^2) = m_3 - 1$
G_2^3	$z(G_2^3) = m_2 - c(G_{23})$ $c(G_2^3) = 1$	$z(G_2^3) = m_2 - 1$	$z(G_2^3) = m_2 - c(G_{12})$ $c(G_2^3) = 1$

(3.3.22) Table of necessary and sufficient F_i-connectivity criteria

Necessary conditions for F_i-connectivity of a three-factor design D can be obtained through the block designs induced by D (for part (a) of the following theorem cf. also Raghavarao/Federer [1975]) :

3.3.23 Theorem

Consider the two-factor designs D_{12} , D_{23} , D_{13} induced by a given three-factor design D . The following holds:

(a) If D is a connected design, then so are D_{23} and D_{13} .

(b) If D is an F_1-connected design, then so are D_{12} and D_{13} .

(c) If D is an F_2-connected design, then so are D_{12} and D_{23} .

Proof:

Part (a) is implied by lemma (3.3.14) together with the equivalences (vii) $\Longleftrightarrow$ (x) and (xiii) $\Longleftrightarrow$ (xvi) of theorem (3.3.19) . In analogy to lemma (3.3.14) one can prove:

(3.3.24) If D is an F_1-connected design, then G_{B_2} and G_{B_3} are connected graphs.

(3.3.25) If D is an F_2-connected design, then $G_{B'_1}$ and $G_{B'_2}$ are connected graphs.

(i) $\Longleftrightarrow$ (iii) and (xiii) $\Longleftrightarrow$ (xv) of theorem (3.3.19) together with (3.3.24) imply assertion (b) . Finally, (c) follows from (i) $\Longleftrightarrow$ (iv) and (vii) $\Longleftrightarrow$ (ix) of theorem (3.3.19) together with (3.3.25) .

□

The next theorem yields a characterization of completely connected three-factor designs D (cf. definition (2.1.13)) by means of the connectivity of D and the connectivity of its induced block designs. Some of the results comprising this theorem are also partially contained in Raghavarao/Federer [1975] , although their proof is based on a different approach.

3.3.26 Theorem

Consider a three-factor design D and its induced two-factor designs D_{12} , D_{23} , D_{13} . The following conditions are equivalent:

(i) D is a completely connected design ,

(ii) D and D_{12} are connected designs ,

(iii) D is F_1-connected and D_{12} is connected ,

(iv) D is F_2-connected and D_{13} is connected .

Proof:

By theorem (2.3.12) the design D is connected if and only if

$$\text{rank } B = m_1 + m_2 + m_3 - 2 \quad .$$

Condition (ii) implies (by lemma (3.1.2))

$$\text{rank } (B_2, B_3) = m_3 - 1 + \text{rank } B_2$$

and (by theorem (3.3.19) and lemma (3.1.3))

$$\text{rank } B_2 = m_2 - 1 \ .$$

On the other hand we have

$$\text{rank } B = m_1 + \text{rank } (B_2, B_3)$$

(cf. (3.3.7)), hence with the above (i) is implied by (ii).

In analogy to lemma (3.1.2) one can prove that D is an F_1-connected design if and only if

$$\text{rank } (B_1', B_3') = m_1 - 1 + \text{rank } B_3' \ ,$$

and an F_2-connected design if and only if

$$\text{rank } (B_1'', B_2'') = m_2 - 1 + \text{rank } B_1'' \ .$$

Therefore condition (iii) implies (by theorem (3.3.19) and lemma (3.1.3))

$$\begin{aligned} \text{rank } B &= m_2 + \text{rank } (B_1', B_3') \\ &= m_2 + m_1 - 1 + \text{rank } B_3' \\ &= m_2 + m_1 - 1 + m_3 - 1 \ , \end{aligned}$$

i. e. (iii) implies (i).

Similarly one can derive condition (i) from condition (iv) :

$$\text{rank } B = m_3 + \text{rank } (B_1'', B_2'')$$

$$= m_3 + m_2 - 1 + \text{rank } B_1''$$

$$= m_3 + m_2 - 1 + m_1 - 1$$

Conversely (ii) , (iii) and (iv) follow from (i) by theorem (3.3.23). □

Theorem (3.3.26) can be illustrated by a set diagram: Let M_{ii^*} $(1 \leq i < i^* \leq 3)$ denote the set of three-factor designs D for which the induced block design D_{ii^*} is connected, and let M_i $(1 \leq i \leq 3)$ denote the set of F_i-connected and M_o the set of completely connected three-factor designs. By theorem (3.3.26) we have

$$M_1 \subseteq M_{12} \cap M_{13} \quad ,$$

$$M_2 \subseteq M_{12} \cap M_{23} \quad ,$$

$$M_3 \subseteq M_{13} \cap M_{23} \quad ,$$

and

$$M_o = M_1 \cap M_2 \cap M_3$$

$$= M_1 \cap M_2 = M_2 \cap M_3 = M_1 \cap M_3$$

$$\subseteq M_{12} \cap M_{23} \cap M_{13} \quad .$$

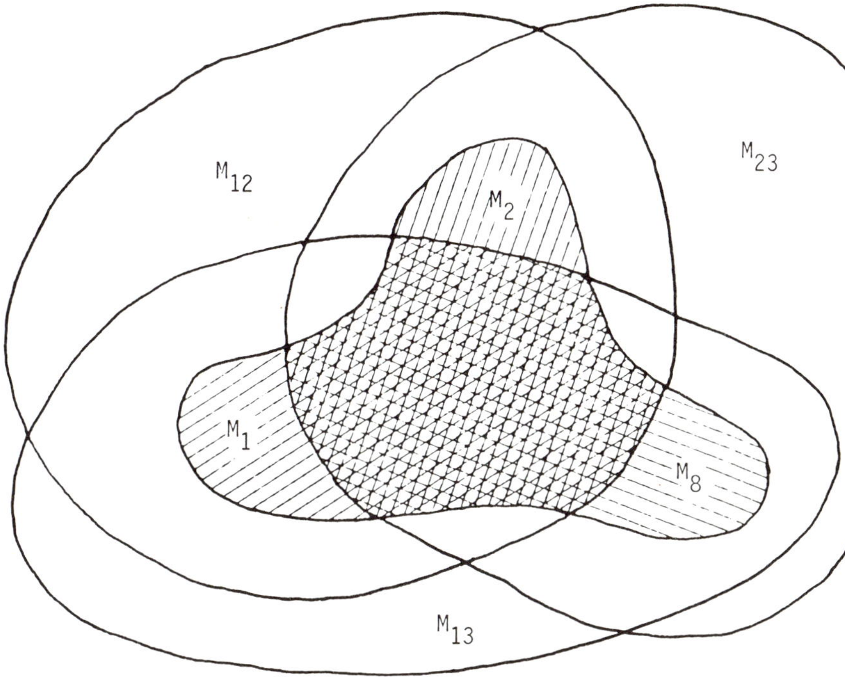

To end this section we shall investigate how the results on the estimability of all F_i-contrasts of a three-factor design obtained above can be used for the analysis of more general classes of linear functions in the parameters of the model (2.1.1) . Let

$$L = \{f \mid f^T = ((f^1)^T,(f^2)^T,(f^3)^T),\ f^i \quad \mathbb{R}^{m_i} \text{ for } i = 1,2,3 ,$$

$$\sum_{j=1}^{m_1} f_j^1 = \sum_{j=1}^{m_2} f_j^2 = \sum_{j=1}^{m_3} f_j^3 \} ,$$

by lemma (2.1.10) it follows that if $\varphi(\pi) = f^T\pi$ is an estimable function in the parameters of the model (2.1.1) , where B defines a three-factor design D , then f is an element of the set L . Now let $L_i \subseteq L$ be the set of all $f \in L$ for which $\varphi(\pi) = f^T(\pi)$ is an F_i-contrast, i. e.

$$L_i = \{f \mid f \in L \, , \, f^\nu = 0_{m_\nu} \text{ for } \nu \in \{1,2,3\} \setminus \{i\}\} \, , \, i = 1,2,3 \, ,$$

and let

$$\tilde{L}_i = \{f \mid f \in L \, , \, f^i = 0_{m_i}\} \, , \, i = 1,2,3 \, ,$$

$$L^\lambda = \{f \mid f \in L \, , \, \sum_{j=1}^{m_\nu} f_j^\nu = \lambda \text{ for } \nu = 1,2,3\} \, , \, \lambda \in \mathbb{R} \, ,$$

$$\hat{L} = \{f \mid f \in L \, , \, f^\nu = u_{m_\nu}^{j_\nu} \, , \, 1 \leq j_\nu \leq m_\nu \text{ for } \nu = 1,2,3\} \, ,$$

where $u_{m_\nu}^{j_\nu}$ denotes the j_ν-th unit vector of $\mathbb{R}^{m_\nu}$. Estimability of a function $\varphi: \pi \longmapsto f^T\pi$ is defined with respect to the underlying linear model $x = B\,\pi + \varepsilon$ (cf. (2.1.1)): $\varphi(\pi) = f^T\pi$ is estimable if and only if $\varphi(\pi) = w^T B\,\pi$ for some $w \in \mathbb{R}^p$. We shall use the formulation "y is estimable for D" whenever we wish to emphasize that the linear model under consideration is determined by a given design D . Finally , for the sake of brevity, let

$$M(X) \qquad (X \subseteq \mathbb{R}^{m_1+m_2+m_3})$$

denote the set of all three-factor designs D for which the function $\varphi(\pi) = f^T\pi$ is estimable for any $f \in X$. Using this notation we can now state the last theorem of this section:

3.3.27 Theorem

Consider a three-factor design D . The following conditions are equivalent:

(i) $D \in M(L)$

(ii) $D \in M(L^1)$

(iii) $D \in M(L^o)$

(iv) $D \in M(\tilde{L}_1)$

(v) $D \in M(\tilde{L}_2)$

(vi) $D \in M(\tilde{L}_3)$

(vii) $D \in M(L_1)$ and G_{23} is a connected graph .

(viii) $D \in M(L_2)$ and G_{13} is a connected graph .

(ix) $D \in M(L_3)$ and G_{12} is a connected graph .

(x) $D \in M(L_1 \cup L_2)$

(xi) $D \in M(L_1 \cup L_3)$

(xii) $D \in M(L_2 \cup L_3)$

(xiii) $D \in M(\hat{L})$

(xiv) D is a completely connected design .

Proof:

By lemma (2.1.9) we know that

$$D \in M(X) \iff (\forall\, f \in X \quad B\,\pi = 0 \implies f^T\pi = 0) \quad ,$$

where B is the model matrix (cf. (2.1.1)) associated with the design D . This implies the following inclusions:

$$M(L) \subseteq M(L^1) \quad ,$$

$$M(L) \subseteq M(L^o) \quad ,$$

$$M(L^1) \subseteq M(\hat{L}) \quad ,$$

$$M(L^o) \subseteq M(\tilde{L}_i) \quad , \qquad i \in \{1,2,3\} \quad ,$$

$$M(\tilde{L}_i) \subseteq M(L_{i'} \cup L_{i''}) \quad , \qquad \{i,i',i''\} = \{1,2,3\} \quad ,$$

$$M(L_i \cup L_{i'}) \subseteq M(L_i) \quad , \qquad i,i' \in \{1,2,3\} \quad ,$$

$$M(L_1 \cup L_2 \cup L_3) \subseteq M(L_i \cup L_{i'}) \ , \ i,i' \in \{1,2,3\} \quad ,$$

$$M(L^o) \subseteq M(L_1 \cup L_2 \cup L_3) \quad .$$

D is a completely connected design if and only if $D \in M(L_1 \cup L_2 \cup L_3)$. Hence, by theorems (3.3.23) and (3.3.26) , each of the conditions (vii) through (xii) is equivalent to condition (xiv) . In the diagram below the solid arrows indicate that the corresponding implications follow from the above set inclusions. Equivalent conditions are listedone under the other.

(3.3.28)

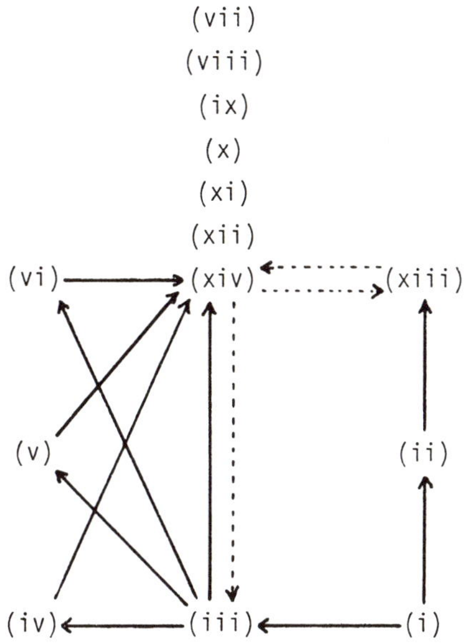

We will now verify the implications which correspond to the dotted arrows.

(xiv) $\Longrightarrow$ (iii) :

For a completely connected D we have

$$\text{kernel } B = \{\pi \mid \exists \lambda, \mu \in \mathbb{R} \ \ \pi^1 = \lambda \cdot 1_{m_1}, \ \pi^2 = \mu \cdot 1_{m_2}, \ \pi^3 = -(\lambda+\mu) \cdot 1_{m_3}\},$$

cf. the proof of theorem (2.3.12) . Hence for all $f \in L^0$

$$B\pi = 0 \quad \Longrightarrow \quad f^T \pi = 0 ,$$

i. e. $D \in M(L^0)$.

(xiii) $\Longleftrightarrow$ (xiv) :

Suppose D is completely connected. Then for each $\pi \in$ kernel B and for all $f \in \hat{L}$

$$f^T\pi = \lambda + \mu - (\lambda + \mu)$$

$$= 0 \quad ,$$

hence $D \in M(\hat{L})$. Now let $D \in M(\hat{L})$, but suppose there exists a $\pi \in$ kernel B with

$$\pi_j^i \neq \pi_{j^*}^i \qquad \text{for some} \quad i \in \{1,2,3\} \; , \quad 1 \leq j < j^* \leq m_i \quad .$$

Consider $\bar{f},\tilde{f} \in \hat{L}$ such that

$$\bar{f}^\nu = \tilde{f}^\nu \qquad \text{for} \quad \nu \neq i \; , \; \nu \in \{1,2,3\} \quad ,$$

$$\bar{f}_k^i = \begin{cases} 1 & \text{for} \quad k = j \quad , \\ 0 & \text{otherwise} \quad , \end{cases}$$

$$\tilde{f}_k^i = \begin{cases} 1 & \text{for} \quad k = j^* \; , \\ 0 & \text{otherwise} \quad . \end{cases}$$

Obviously by construction $\bar{f}^T\pi \neq \tilde{f}^T\pi$, hence we cannot have $f^T\pi = 0$ for all $f \in \hat{L}$, which contradicts $D \in M(\hat{L})$. This proves the equivalence of conditions (xiii) and (xiv) .

Now consider a further condition:

(xv) $\qquad D \in M(L^o \cup L^1)$.

As $M(L^o)$ and $M(L^1)$ both are subsets of $M(L^o \cup L^1)$ we can draw another diagram

(3.3.29)

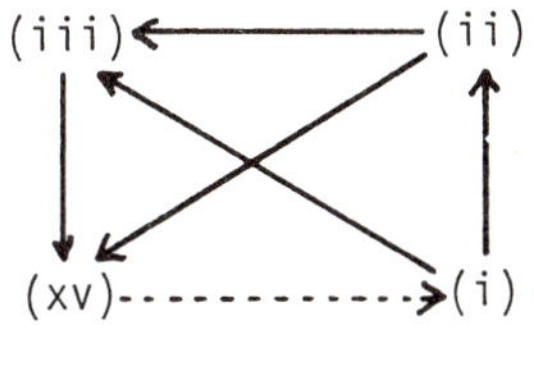

where again implications are indicated by solid arrows. Here the implication (ii) $\Longrightarrow$ (iii) follows from (ii) $\Longrightarrow$ (xiii) and from the equivalence of condition (iii) to each of the conditions (iv) through (xiv) , cf. (3.3.28) . To complete the proof of the theorem it suffices to verify

(xv) $\Longrightarrow$ (i) :

Suppose for each $f \in L^o \cup L^1$ we have

$$B\pi = 0 \Longrightarrow f^T\pi = 0 \quad .$$

Choose any fixed $f \in L$. In case $f \in L^o$ nothing remains to be shown; for $f \in L \setminus L^o$ we have

$$f \in L^\lambda \text{ for some positive } \lambda \in \mathbb{R} \quad .$$

Now let $\tilde{f} = \lambda^{-1} \cdot f$. Clearly $\tilde{f} \in L^1$, therefore $f^T\pi = \lambda \cdot \tilde{f}^T\pi = 0$ for all $\pi \in$ kernel B .

$\square$

The preceding theorem offers informative insight into estimability characterizations of completely connected three-factor designs: Take a linear function $\varphi(\pi) = f^T\pi$ which is estimable for D , then by lemma (2.1.10) f is a element of the set L . Conversely, theorem (3.3.27) states that any $f \in L$ yields an estimable function for a completely connected D . Furthermore, we even know that the estimability of special subclasses of linear functions is necessary and sufficient for D to be completely connected .

Moreover, it should be pointed out that $D \in M(\hat{L})$ is equivalent to the estimability of

$$\pi_i^1 + \pi_j^2 + \pi_k^3$$

for any $1 \leq i \leq m_1$, $1 \leq j \leq m_2$, $1 \leq k \leq m_3$, i. e. the expectation of the outcome of an experiment with an arbitrary combination of factor levels (including experiments which are not even contained in the experimental design).

3.4 Four-factor designs

A four-factor design is determined by a nonnegative integer $m_1 \times m_2 \times m_3 \times m_4$ matrix D . Also for these designs one can introduce a digraph with labeled arcs which yields a necessary and sufficient characterization of connectivity. For this purpose let $Y = ((y_{j_1j_2j_3}))$ denote an $m_1 \times m_2 \times m_3$ matrix of sets

$$y_{j_1j_2j_3} = \{k \mid d_{j_1j_2j_3k} > 0\}$$

$$\text{where} \quad j_i = 1,\dots,m_i \quad , \quad i = 1,2,3 \quad .$$

Let J be the set of all ordered index pairs (j_1,j_2) for which $d_{j_1j_2 1k} = 0$ holds for each treatment k , i. e.

$$J = \{(j_1,j_2) \mid \sum_{k=1}^{m_4} d_{j_1j_2 1k} = 0\}$$

$$= \{(j_{1,\nu},j_{2,\nu}) \mid \nu = 1,\dots,s\}$$

where $s = |J|$ and $(j_{1,\nu},j_{2,\nu})$ is lexicographically smaller than $(j_{1,\nu+1},j_{2,\nu+1})$ for $\nu = 1,\dots,s-1$. The elements of J correspond to certain empty entries of the set matrix Y , and the digraph to be defined will contain special vertices for these empty entries which have the sole purpose of making possible a very simple labeling of the arcs in the resulting digraph. Given Y and J we take

$$m_{j_1j_2j_3} = \begin{cases} \min\{k \mid k \in y_{j_1j_2j_3}\} & \text{if } y_{j_1j_2j_3} \neq \emptyset , \\ m_4 + \nu & \text{if } y_{j_1j_2j_3} = \emptyset , \; j_3 = 1 , \\ & \quad (j_1,j_2) = (j_{1,\nu},j_{2,\nu}) \in J , \\ 0 & \text{if } y_{j_1j_2j_3} = \emptyset , \; j_3 > 1 , \end{cases}$$

$$\text{for } j_i = 1,\dots,m_i , \quad i = 1,2,3 .$$

We can now associate the following digraph G with the four-factor design D .

$$G = (V,E) ,$$

$$V = \{1,\dots,m_4+s\} ,$$

$$E = E_1 \cup E_2 \cup E_3 ,$$

$$E_1 = \{(k,k^*)_0 \mid \exists\, j_1,j_2,j_3 \text{ with } k^* = m_{j_1j_2j_3} < k \in y_{j_1j_2j_3}\} ,$$

$$E_2 = \{(k,k^*)_j \mid \exists\, j_1 \text{ with } k = m_{j_1j1} > 0 , \; k^* = m_{j_111} , \; j \geq 2\} ,$$

$$E_3 = \{(k,k^*)_j \mid \exists\, j_1,j_2,j_3 \text{ with } k = m_{j_1j_2j_3} > 0 , \; k^* = m_{j_1j_21} , \; j = m_2 + j_3 , \; j_3 \geq 2\} .$$

The construction of G is easily illustrated by an example.

3.4.1 Example

Consider a design given by a $4 \times 3 \times 2 \times 8$ matrix D with

$d_{1111} = 1$, $d_{1112} = 1$, $d_{1214} = 1$, $d_{1313} = 1$,

$d_{1315} = 1$, $d_{2214} = 1$, $d_{2218} = 1$, $d_{2316} = 1$,

$d_{3112} = 1$, $d_{3117} = 1$, $d_{3211} = 1$, $d_{3312} = 1$,

$d_{4116} = 1$, $d_{4213} = 1$, $d_{4314} = 1$, $d_{4318} = 1$,

$d_{1227} = 1$, $d_{1325} = 1$, $d_{1326} = 1$, $d_{2125} = 1$,

$d_{3127} = 1$, $d_{3324} = 1$, $d_{3328} = 1$, $d_{4224} = 1$,

$d_{j_1 j_2 j_3 j_4} = 0$ otherwise .

The $4 \times 3 \times 2$ set matrix Y can be stated in the form

$$Y_{--1} = \begin{bmatrix} 1,2 & 4 & 3,5 \\ - & 4,8 & 6 \\ 2,7 & 1 & 2 \\ 6 & 3 & 4,8 \end{bmatrix} , \quad Y_{--2} = \begin{bmatrix} - & 7 & 5,6 \\ 5 & - & - \\ 7 & - & 4,8 \\ - & 4 & - \end{bmatrix} ,$$

hence it follows that for $((m_{j_1 j_2 j_3}))$ one gets

$$M_{--1} - \begin{bmatrix} 1 & 4 & 3 \\ 9 & 4 & 6 \\ 2 & 1 & 2 \\ 6 & 3 & 4 \end{bmatrix} , \quad M_{--2} = \begin{bmatrix} 0 & 7 & 5 \\ 5 & 0 & 0 \\ 7 & 0 & 4 \\ 0 & 4 & 0 \end{bmatrix} ,$$

where clearly $J = \{(2,1)\}$ and $s = |J| = 1$. The associated digraph G is

G :

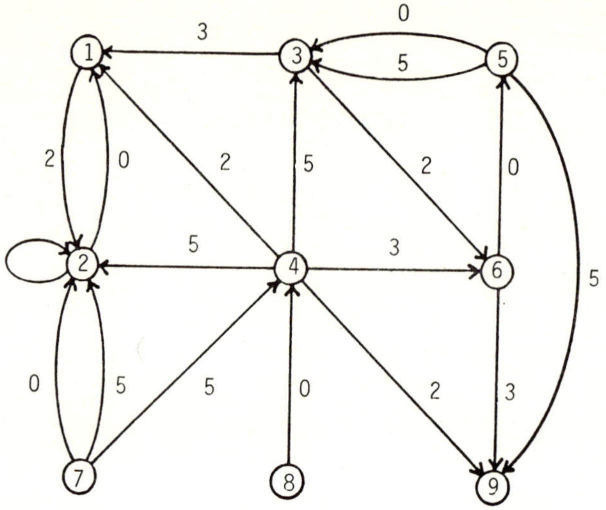

.

For each $y_{j_1j_2j_3}$ with $|y_{j_1j_2j_3}| \geqq 2$ the digraph G contains exactly $|y_{j_1j_2j_3}| - 1$ arcs which are labeled 0 ,

$$(k,m_{j_1j_2j_3})_0 \ , \ k \in y_{j_1j_2j_3} \setminus \{m_{j_1j_2j_3}\} .$$

The remaining arcs are of the form

$$(m_{j_1j1},m_{j_111})_j \ , \ j \geqq 2 \ ,$$

or

$$(m_{j_1j_2j_3},m_{j_1j_21})_{m_2+j_3} \ , \ j_3 \geqq 2 \ ,$$

their labels resulting from the corresponding indices of the tail vertices.

Consider the three-factor design

$$D_{123} = ((d^{123}_{j_1j_2j_3}))$$

which is induced by a given four-factor design D by means of

$$d^{123}_{j_1j_2j_3} = \sum_{j=1}^{m_4} d_{j_1j_2j_3j} \quad , \quad j_i = 1,\dots,m_i \quad , \quad i = 1,2,3 \quad ,$$

and its associated digraph G^3_2 (cf. section 3.3 and the digraph defined for theorem (3.3.5)) . The matrix B of (2.1.1) can be brought into the form

$$\text{(3.4.2)} \qquad PAB = \begin{pmatrix} I_{m_1} & \tilde{B}_2 & \tilde{B}_3 & \tilde{B}_4 \\ 0_{p-m_1,m_1} & B_2 & B_3 & B_4 \end{pmatrix} ,$$

by elementary row operations, cf. (3.1.1). Lemma (3.1.2) says that D is a connected design if and only if the submatrices B_2 , B_3 , B_4 of (3.4.2) satisfy

$$\operatorname{rank}(B_2,B_3,B_4) = m_4 - 1 + \operatorname{rank}(B_2,B_3) \quad .$$

The following lemma shows the relevance of the induced design D_{123} for the connectivity of D .

<u>3.4.3 Lemma</u>

Consider a four-factor design D , its induced design D_{123} , and the digraph G^3_2 associated with D_{123} . Then the submatrices B_2,B_3 of (3.4.2) satisfy

$$\operatorname{rank}(B_2,B_3) = m_3 - c(G^3_2) + z(G^3_2) \quad .$$

Proof:

Let the coefficient matrix of the model (2.1.1) for the three-factor design D_{123} be denoted by B^*. To avoid confusion with the matrices appearing in (3.4.2) we represent equation (3.3.7) as

$$P^* A^* B^* = \begin{pmatrix} I_{m_1} & \tilde{B}_2^* & \tilde{B}_3^* \\ 0_{p^*-m_1,m_1} & B_2^* & B_3^* \end{pmatrix} ,$$

Now it is easy to see that B^* is just a submatrix of the matrix B (which corresponds to the four-factor design D) , namely the submatrix obtained by deleting the last m_4 columns of B . Furthermore it also follows immediately that $p = p^*$, $P = P^*$, $A = A^*$, $\tilde{B}_2 = \tilde{B}_2^*$, $\tilde{B}_3 = \tilde{B}_3^*$, and in particular that $B_2 = B_2^*$ and $B_3 = B_3^*$.

Lemma (3.1.3) yields for the digraph G_{B_2,B_3} :

$$c(G_{B_2,B_3}) = c(G_2^3) = m_3 - \text{rank } B_3 .$$

For the maximum number of independent cycles in G_{B_2,B_3} we know (cf. the proof of theorem (3.3.5)) that

$$z(G_{B_2,B_3}) = z(G_2^3) ,$$

therefore by lemma (3.1.5) it follows that

$$\text{rank } (B_2,B_3) = \text{rank } B_3 + z(G_{B_2,B_3}) = m_3 - c(G_2^3) + z(G_2^3) . \qquad \square$$

For the connectivity criterion to be derived we need vector labels for the cycles in the digraph G associated to a given four-factor design. We therefore define a function g which maps chains τ of G into $\mathbb{Z}^{m_2+m_3}$ by

$$g(\tau) = \sum_{e_j \in E(\tau^+)} u^j - \sum_{e_j \in E(\tau^-)} u^j ,$$

where u^j is the j-th unit vector of $\mathbb{R}^{m_2+m_3}$ and $u^0 = 0_{m_2+m_3}$. The only labels of the arcs in the set E that appear in G are $2,\dots,m_2$ and $m_2+2,\dots,m_2+m_3$, hence the first and (m_2+1)-th component of each vector label $g(\tau)$ vanishes.

<u>3.4.4 Theorem</u>

For a given four-factor design D consider the associated digraph G , the induced three-factor design D_{123} and the digraph G_2^3 given by D_{123} .Then the following holds: D is a connected design if and only if G is a connected digraph which contains exactly $m_3 - c(G_2^3) + z(G_2^3)$ independent cycles.

<u>Proof:</u>

The digraph $G_{(B_2,B_3),B_4}$ is connected and has

$$\text{rank } (B_2,B_3) = m_3 - c(G_2^3) + z(G_2^3)$$

independent cycles if and only if D is a connected design (cf. lemmas (3.1.2) , (3.1.5) taking $R = (B_2,B_3)$, $Q = B_4$, and lemma (3.4.3)) . We wish to show that for each elementary cycle $\hat{\sigma}$ in $G_{(B_2,B_3),B_4}$ there is a cycle σ in G such that

(3.4.5) $$h(\hat{\sigma}) = \text{diag}\,(0\,,1^T_{m_2-1}\,,0\,,1^T_{m_3-1})\;g(\sigma)\quad,$$

i. e. $h(\hat{\sigma})$ and $g(\sigma)$ differ in at most the first and (m_2+1)-th components.

Consider an arc $(k',k")_r$ contained in an elementary cycle $\hat{\sigma}$ of $G_{(B_2,B_3),B_4}$, where r denotes a row vector of the matrix (B_2,B_3) :

$$r = ((u^{j''_2}_{m_2} - u^{j'_2}_{m_2})^T\,,\,(u^{j''_3}_{m_3} - u^{j'_3}_{m_3})^T)\quad.$$

The row of (B_2,B_3,B_4) which corresponds to the arc $(k',k")_r$ is of the form $(r\,,(u^{k"}_{m_4} - u^{k'}_{m_4})^T)$ by construction of $G_{(B_2,B_3),B_4}$. Hence there exist $\tilde{k}\,,\,\hat{k} \in \{1,\dots,m_4\}$, $j^*_1 \in \{1,\dots,m_1\}$ such that

$$d_{j^*_1j'_2j'_3\tilde{k}} > 0\quad,\quad d_{j^*_1j''_2j''_3\hat{k}} > 0\quad,$$

where $(j^*_2,j''_3,\hat{k})$ is the lexicographically smallest triple (j_2,j_3,j_4) with $d_{j^*_1j_2j_3j_4} > 0$. Here we have $\tilde{k} = k'$, $\hat{k} = k"$ in case $k' \neq k"$, and $\tilde{k} = \hat{k}$ in case $k' = k"$ (cf. construction of P A B in section 3.1) .

Consider for example the $m_2 \times m_3$ matrix $M_{j^*_1--} = ((m_{j^*_1j_2j_3}))$:

$$M_{j_1^*--} = \begin{bmatrix} m_{j_1^*11} & \cdots & m_{j_1^*1j_3''} & \cdots & m_{j_1^*1j_3'} & \cdots \\ m_{j_1^*j_2''1} & \cdots & \boxed{\hat{k} = m_{j_1^*j_2''j_3''}} & \cdots & \vdots & \cdots \\ \vdots & & \vdots & & \vdots & \\ m_{j_1^*j_2'1} & \cdots & & \cdots & \boxed{m_{j_1^*j_2'j_3'}} & \cdots \\ \vdots & & \vdots & & \vdots & \end{bmatrix}$$

Here $1 \leq j_2'' \leq j_2' \leq m_2$ and $m_{j_1^*j_2'j_3'} \leq \tilde{k} \in y_{j_1^*j_2'j_3'}$. For the above representation of $M_{j_1^*--}$ we have assumed the case $1 < j_2'' < j_2'$, $m_{j_1^*j_2'j_3'} < \tilde{k}$, $1 < j_3''$, $1 < j_3'$. If any of the conditions $j_2'' = 1$, $j_2' = j_2''$, $m_{j_1^*j_2'j_3'} = \tilde{k}$, $j_3' = 1$, $j_3'' = 1$ hold, simple special cases result. By definition of the digraph $G = (V,E)$ it is easy to see that

$$(\tilde{k}\,, m_{j_1^*j_2'j_3'})_0 \in E_1 \quad ,$$

$$(m_{j_1^*j_2'1}\,, m_{j_1^*11})_{j_2'} \quad , \quad (m_{j_1^*j_2''1}\,, m_{j_1^*11})_{j_2''} \in E_2 \quad ,$$

$$(\hat{k}\,, m_{j_1^*j_2''1})_{m_2+j_3''} \quad , \quad (m_{j_1^*j_2'j_3'}\,, m_{j_1^*j_2'1})_{m_2+j_3'} \in E_3 \quad .$$

Therefore the chain

τ :

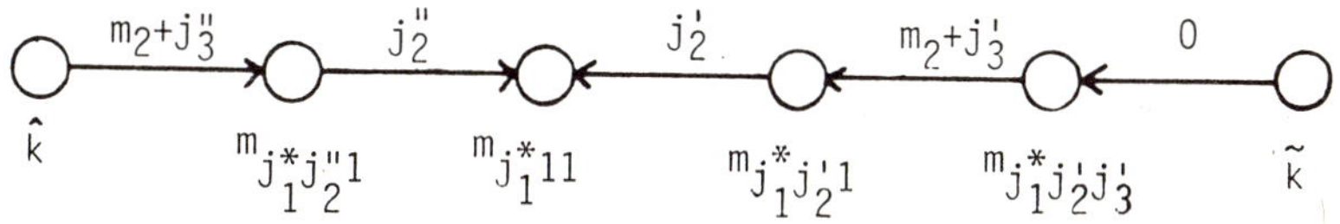

is contained in G . Furthermore $\hat{\tau} = [k', (k',k'')_r, k'']$ is a chain in $G_{(B_2,B_3),B_4}$ and we have

$$g(\tau) = u_{m_2+m_3}^{m_2+j_3''} + u_{m_2+m_3}^{j_2''} - u_{m_2+m_3}^{j_2'} - u_{m_2+m_3}^{m_2+j_3'}$$

$$= r^T = h(\hat{\tau}) \quad .$$

An investigation of the cases $j_2'' = 1$, $j_3'' = 1$ or $j_3' = 1$ shows that $g(\tau)$ and $h(\hat{\tau})$ can only differ in the first and (m_2+1)-th component (cf. the cases treated in the proof of theorem (3.3.5)) . Now an easy induction on the number of arcs in $\hat{\sigma}$ yields the desired result.

Conversely one can also show that for each chain σ in G with tail vertex k' $(1 \leq k' \leq m_3)$ and head vertex k'' $(1 \leq k'' \leq m_3)$, there is a chain $\hat{\sigma}$ in $G_{(B_2,B_3),B_4}$ with tail vertex k'' and head vertex k' such that (3.4.5) holds.

Every cycle in G contains at least one vertex $k \in V$ with $1 \leq k \leq m_3$. Furthermore it is easy to see that the maximum number of independent elementary cycles in G is equal to the maximum number of independent cycles in G . By definition of the labeling function h we know that for each chain $\hat{\tau}$ in $G_{(B_2,B_3),B_4}$

$$(1_{m_2}^T , 0_{m_3}^T) \, h(\hat{\tau}) = 0 = (0_{m_2}^T , 1_{m_3}^T) \, h(\hat{\tau}) \quad .$$

As a result the following equivalence holds:

$G_{(B_2,B_3),B_4}$ is a connected digraph having exactly ℓ independent cycles

$\Longleftrightarrow$ G is a connected digraph having exactly ℓ independent cycles.

□

For the four-factor design D treated in example (3.4.1) the induced three-factor design D_{123} is given by

$$d^{123}_{111} = 2\ ,\ d^{123}_{121} = 1\ ,\ d^{123}_{122} = 1\ ,\ d^{123}_{131} = 2\ ,$$

$$d^{123}_{132} = 2\ ,\ d^{123}_{212} = 1\ ,\ d^{123}_{221} = 2\ ,\ d^{123}_{231} = 1\ ,$$

$$d^{123}_{311} = 2\ ,\ d^{123}_{312} = 1\ ,\ d^{123}_{321} = 1\ ,\ d^{123}_{331} = 1\ ,$$

$$d^{123}_{332} = 2\ ,\ d^{123}_{411} = 2\ ,\ d^{123}_{421} = 1\ ,\ d^{123}_{422} = 2\ ,$$

$$d^{123}_{431} = 2 \text{ and } d^{123}_{j_1j_2j_3} = 0 \text{ otherwise }.$$

The sets $y_{ij} = \{d^{123}_{ijk} > 0\}$ can be read off as

$$((y_{ij})) = \begin{bmatrix} 1 & 1,2 & 1,2 \\ 2 & 1 & 1 \\ 1,2 & 1 & 1,2 \\ 1 & 1,2 & 1 \end{bmatrix} ,$$

hence the associated digraph G^3_2 is

G_2^3 :

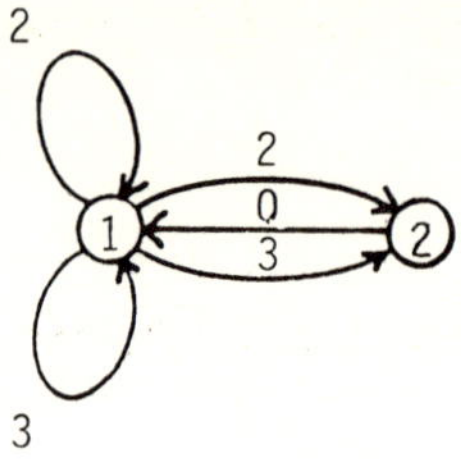

G_2^3 is connected and the cycles

$$\sigma = [1\,,(1,1)_2\,,1] \quad , \quad \sigma' = [1\,,(1,1)_3\,,1]$$

clearly have independent vector labels. Therefore D_{123} is a completely connected design,

$$m_3 - c(G_2^3) + z(G_2^3) = 2 - 1 + 2 = 3 \quad .$$

By theorem (3.4.4) the design D is connected if and only if the digraph G shown under (3.4.1) is connected and contains 3 independent cycles . These conditions are fulfilled: Consider for instance the cycles

$$\sigma_1 = [1\,,(1,2)_0\,,2\,,(2,1)_2\,,1] \quad ,$$

$$\sigma_2 = [5\,,(5,3)_3\,,3\,,(5,3)_0\,,5] \quad ,$$

$$\sigma_3 = [7\,,(7,2)_5\,,2\,,(7,2)_0\,,7] \quad ,$$

and their vector labels

$$g(\sigma_1) = (0,1,0,0,0)^T \quad ,$$

$$g(\sigma_2) = (0,0,1,0,0)^T \quad ,$$

$$g(\sigma_3) = (0,0,0,0,1)^T \quad .$$

The criterion of theorem (3.4.4) becomes a great deal simpler for four-factor designs satisfying

$$\sum_{j=1}^{m_4} d_{j_1 j_2 j_3 j} = 1 \quad , \quad j_i = 1,\dots,m_i \quad , \quad i = 1,2,3 \quad ,$$

which is a case that occurs quite frequently. In analogy to the row-column designs treated in section 3.2 one can represent such a design by an $m_1 \times m_2 \times m_3$ matrix $Y = ((y_{j_1 j_2 j_3}))$ with entries $y_{j_1 j_2 j_3} \in \{1,\dots,m_4\}$, where

$$y_{j_1 j_2 j_3} = j \iff d_{j_1 j_2 j_3 j} > 0 \quad .$$

The associated digraph $G = (V,E)$ of theorem (3.4.4) can be seen to have the vertex set

$$V = \{1,\dots,m_4\}$$

and the arc set

$$E = \{(y_{j_1 j 1}, y_{j_1 1 1})_j \mid 1 \leq j_1 \leq m_1 \ , \ 2 \leq j \leq m_2\}$$

$$\cup \ \{(y_{j_1 j_2 j}, y_{j_1 j_2 1})_{m_2+j} \mid 1 \leq j_1 \leq m_1, \ 1 \leq j_2 \leq m_2 \ , \ 2 \leq j \leq m_3\}.$$

3.4.6 Corollary

Consider a four-factor design D in which each combination of blocking factor levels occurs exactly once. D is a connected design if and only if the associated digraph G is connected and the number of independent cycles in G is $m_2 + m_3 - 2$.

Proof:

The digraph G_2^3 corresponding to the three-factor design D_{123} induced by D has the following form:

G_2^3 :

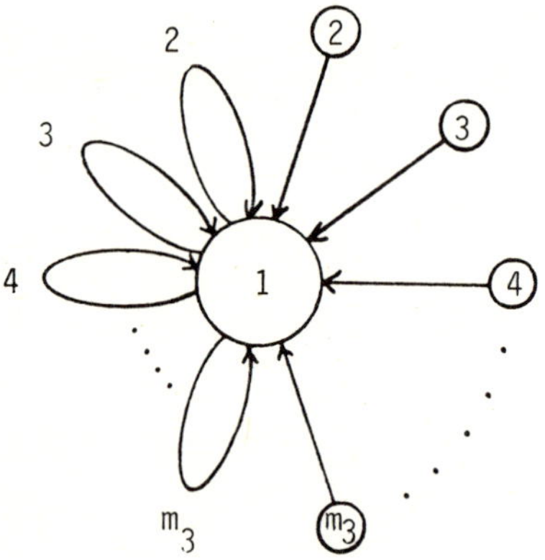

,

Therefore D_{123} is completely connected and

$$m_3 - c(G_2^3) + z(G_2^3) = m_3 - 1 + m_2 - 1 \quad ,$$

which shows that the corollary is an immediately consequence of theorem (3.4.4).

□

It is desirable to minimize the effort required to check a connectivity criterion, and for this reason we will outline some modifications of the characterization in theorem (3.4.4) which will provide "easy" criteria depending on the relative sizes of the numbers m_1, m_2, m_3, m_4 .

For this purpose we transform the model matrix B (cf. (2.1.1)) of a given four-factor design D in four different ways by means of elementary row operations. Multiplication from the left with appropriate non-singular $p \times p$ matrices Z^i (i = 1,2,3,4) yields

$$(3.4.7) \qquad Z^1 B = \begin{bmatrix} I_{m_1} & \tilde{B}_2 & \tilde{B}_3 & \tilde{B}_4 \\ 0_{p-m_1,m_1} & B_2 & B_3 & B_4 \end{bmatrix} ,$$

$$(3.4.8) \qquad Z^2 B = \begin{bmatrix} \tilde{B}_1^2 & I_{m_2} & \tilde{B}_3^2 & \tilde{B}_4^2 \\ B_1^2 & 0_{p-m_2,m_2} & B_3^2 & B_4^2 \end{bmatrix} ,$$

$$(3.4.9) \qquad Z^3 B = \begin{bmatrix} \tilde{B}_1^3 & \tilde{B}_2^3 & I_{m_3} & \tilde{B}_4^3 \\ B_1^3 & B_2^3 & 0_{p-m_3,m_3} & B_4^3 \end{bmatrix} ,$$

$$(3.4.10) \qquad Z^4 B = \begin{bmatrix} \tilde{B}_1^4 & \tilde{B}_2^4 & \tilde{B}_3^4 & I_{m_4} \\ B_1^4 & B_2^4 & B_3^4 & 0_{p-m_4,m_4} \end{bmatrix} .$$

Here (3.4.7) is only a repetition of (3.4.2) , i. e. $Z^1 = PA$. The matrices Z^2 , Z^3 and Z^4 can also be specified easily, however for brevity we will not do so (cf. section 3.3 and remarks in connection with the multipliers $P'A'\overline{P}$ and $P''A''\hat{P}$ of (3.3.8) and (3.3.9)) .

Loosely speaking the digraph G defined for theorem (3.4.4) can be constructed from $G_{(B_2,B_3),B_4}$ by taking an additional set of vertices, namely $\{m_4 + \nu \mid \nu = 1,\ldots,|J|\}$, and diverting certain arcs via these new vertices such that the resulting set of arcs has integral labels, as contrasted with the arcs of $G_{(B_2,B_3),B_4}$ which are labeled with row vectors of (B_2,B_3) (cf. the proof of theorem (3.4.4)). For this reason we will now also denote the digraph G by

$$G_{23}^{4}$$

in order to point out its relation to the digraph $G_{(B_2,B_3),B_4}$ and the role of the factors F^4 ($\longrightarrow$ vertices) as well as F_2 and F_3 ($\longrightarrow$ arc labels) for the definition of G . Obvious variants of G_{23}^{4} can be obtained by considering analogously defined digraphs G_{jk}^{i} where the factor indices $i,j,k \in \{1,2,3,4\}$ are taken pairwise distinct. That is, G_{jk}^{i} results from G_{23}^{4} by interchanging the roles of the factors F_1 , F_2 , F_3 , F_4 according to the permutation

$$\begin{pmatrix} 1 & 2 & 3 & 4 \\ \nu & j & k & i \end{pmatrix}, \text{ where } \{\nu\} = \{1,2,3,4\} \setminus \{i,j,k\} \quad .$$

As the above loose description seems sufficiently explicit, we shall not give a detailed definition of all the G_{jk}^{i} . Denoting the relation between the definitions of the digraphs $G = G_{23}^{4}$ and $G_{(B_2,B_3),B_4}$ by "$\hat{=}$" , we obtain

$$G^i_{jk} \hat{=} G_{(B_j,B_k),B_i} \quad \text{for pairwise distinct} \quad i,j,k \in \{2,3,4\} \quad ,$$

$$G^i_{jk} \hat{=} G_{(B^\nu_j,B^\nu_k),B^\nu_i} \quad \text{for pairwise distinct} \quad i,j,k \in \{1,2,3,4\} \setminus \{\nu\} \; , \quad \nu = 2,3,4$$

with the notation of (3.4.7) through (3.4.10) .

3.4.11 Theorem

Let a four-factor design be given by a non-negative integer $m_1 \times m_2 \times m_3 \times m_4$ matrix D , and consider the induced three-factor design D_{123} , the digraphs G^i_{jk} corresponding to D ($i,j,k \in \{1,2,3,4\}$ pairwise distinct), and G^3_2 corresponding to D_{123} . Then the following holds:

D is a connected design if and only if

$$z(G^i_{jk}) - c(G^i_{jk}) = m_1 + m_3 + m_4 - m_i - m_\nu - 1 + z(G^3_2) - c(G^3_2)$$

where $\{\nu\} = \{1,2,3,4\} \setminus \{i,j,k\}$.

Proof:

It follows from lemmas (3.1.3) and (3.1.5) that

$$\begin{aligned}\operatorname{rank} B &= \operatorname{rank} Z^1 B \\ &= m_1 + \operatorname{rank} B_i + z(G^i_{jk}) \\ &= m_1 + m_i - c(G^i_{jk}) + z(G^i_{jk})\end{aligned}$$

for pairwise distinct $i,j,k \in \{2,3,4\}$,

$$\begin{aligned}\operatorname{rank} B &= \operatorname{rank} Z^2 B \\ &= m_2 + \operatorname{rank} B^2_i + z(G^i_{jk}) \\ &= m_2 + m_i - c(G^i_{jk}) + z(G^i_{jk})\end{aligned}$$

for pairwise distinct $i,j,k \in \{1,3,4\}$,

$$\begin{aligned}\operatorname{rank} B &= \operatorname{rank} Z^3 B \\ &= m_3 + \operatorname{rank} B^3_1 + z(G^i_{jk}) \\ &= m_3 + m_i - c(G^i_{jk}) + z(G^i_{jk})\end{aligned}$$

for pairwise distinct $i,j,k \in \{1,2,4\}$,

and

$$\begin{aligned}\operatorname{rank} B &= \operatorname{rank} Z^4 B \\ &= m_4 + \operatorname{rank} B^4_1 + z(G^i_{jk}) \\ &= m_4 + m_i - c(G^i_{jk}) + z(G^i_{jk})\end{aligned}$$

for pairwise distinct $i,j,k \in \{1,2,3\}$.

From theorem (3.4.4) it follows that D is a connected design if and only if

$$\text{rank } B = m_1 + m_4 - 1 + m_3 + z(G_2^3) - c(G_2^3) \quad .$$

Inserting

$$\text{rank } B = m_\nu - m_i - c(G_{jk}^i) + z(G_{jk}^i)$$

completes the proof. □

It is not clear from theorem (3.4.11) in which cases G_{jk}^i necessarily has to be a connected digraph if the associated design D is to be connected. For $i = 4$ this question is answered by the following theorem.

<u>3.4.12 Theorem</u>

Let an $m_1 \times m_2 \times m_3 \times m_4$ matrix D define a four-factor design, and consider the digraph G_{jk}^4 where $j,k \in \{1,2,3\}$, $j \neq k$, and the digraphs G_2^3 and G_j^k associated to the induced three-factor design D_{123} . Then the following conditions are equivalent:

(i) D is a connected design ;

(ii) $c(G_{jk}^4) = 1$ and $z(G_{jk}^4) = m_1 + m_3 - m_\nu + z(G_2^3) - c(G_2^3)$

where $\{\nu\} = \{1,2,3\} \setminus \{j,k\}$;

(iii) $c(G_{jk}^4) = 1$ and $z(G_{jk}^4) = m_k + z(G_j^k) - c(G_j^k)$.

<u>Proof</u>:

By lemma (3.1.2) and the proof of theorem (3.4.4) we know that D is a connected design if and only if

$$\operatorname{rank} B = m_1 + m_4 - 1 + \operatorname{rank}(B_2, B_3)$$

$$= m_1 + m_4 - 1 + m_3 - c(G_2^3) + z(G_2^3) \quad .$$

Analogously we get, using the notation of (3.4.8) and (3.4.9) ,

$$\text{(i)} \iff \operatorname{rank} B = m_2 + m_4 - 1 + \operatorname{rank}(B_1^2, B_3^2)$$

$$\iff \operatorname{rank}(B_1^2, B_3^2) = m_1 - m_2 + m_3 - c(G_2^3) + z(G_2^3)$$

$$\iff \operatorname{rank} B = m_3 + m_4 - 1 + \operatorname{rank}(B_1^3, B_2^3)$$

$$\iff \operatorname{rank}(B_1^3, B_2^3) = m_1 - c(G_2^3) + z(G_2^3) \quad .$$

By construction of G_{jk}^4 it is clear that $c(G_{jk}^4) = c(G_{kj}^4)$ and $z(G_{jk}^4) = z(G_{kj}^4)$, hence

$$\text{(i)} \iff c(G_{23}^4) = 1 \text{ and } z(G_{23}^4) = m_3 - c(G_2^3) + z(G_2^3)$$

$$\iff c(G_{13}^4) = 1 \text{ and } z(G_{13}^4) = m_1 - m_2 + m_3 - c(G_2^3) + z(G_2^3)$$

$$\iff c(G_{12}^4) = 1 \text{ and } z(G_{12}^4) = m_1 - c(G_2^3) + z(G_2^3)$$

$$\iff \text{(ii)} \quad .$$

Theorem (3.4.4) states that

$$\text{(i)} \iff c(G^4_{23}) = 1 \text{ and } z(G^4_{23}) = m_3 - c(G^3_2) + z(G^3_2) \ .$$

By means of symmetry arguments it is easy to verify (cf. (3.4.8) and (3.4.9)) that

$$\text{(i)} \iff c(G^4_{13}) = 1 \text{ and } z(G^4_{13}) = m_3 - c(G^3_1) + z(G^3_1)$$

$$\iff c(G^4_{12}) = 1 \text{ and } z(G^4_{12}) = m_2 - c(G^2_1) + z(G^2_1)$$

$$\iff \text{(ii)} \ .$$

□

Theorem (3.4.11) and (3.4.12) make it possible to choose a convenient connectivity criterion depending on the dimensions of the four-factor design to be examined (i. e. depending on the relative sizes of m_1 , m_2 , m_3 and m_4) . As one prefers to deal with as few cycles as possible in a digraph which is preferably visible at a glance, we provide a rough guideline for the choice of a suitable digraph G^i_{jk} :

choose j,k such that $m_j + m_k = \min \{m_{j'} + m_{k'} \mid j',k' \in \{1,2,3,4\} \ , \ j' \neq k'\}$;

choose i such that $i = \min \{i' \mid i' \in \{1,2,3,4\} \setminus \{j,k\}\}$.

The results on three-factor designs contained in theorems (3.3.23) , (3.3.26) and (3.3.27) carry over to a certain extent to four-factor designs.

However, a detailed discussion of this is omitted here as there seem to be no aspects which are characteristic for four-factor designs (cf. theorems (3.5.3) , (3,5,5) , (3.5.8) and (3.5.10) for the corresponding results about general n-factor designs) .

3.5 Multi-factor designs

In this last section of the third chapter the general case of a design with n factors will be discussed $(n \in \mathbb{N},\ n \geq 2)$. We assume a representation of the design by means of a non-negative integer $m_1 \times \dots \times m_n$ matrix D with $d^i_j > 0$ $(i = 1,\dots,n\ ,\ j = 1,\dots,m_i)$, cf. (2.1.4). As a direct generalization of the digraph G^4_{23} which was defined for four-factor designs, we introduce a digraph denoted by $G^n_{2\dots n-1}$. As a warning it should be mentioned that for a large number of factors n and for large $m_2,\dots,m_n$ the construction of the digraph becomes rather cumbersome. For such cases alternative concepts will be introduced in the next chapter.

$Y = ((y_{j_1 \dots j_{n-1}}))$ is defined to be an $m_1 \times \dots \times m_{n-1}$ matrix of sets

$$y_{j_1 \dots j_{n-1}} = \{k \mid d_{j_1 \dots j_{n-1} k} > 0\}$$

where $j_i = 1,\dots,m_i$ for $i = 1,\dots,n-1$. Furthermore let

$$J = \{(j_1,\dots,j_{n-2}) \mid \sum_{k=1}^{m_n} d_{j_1 \dots j_{n-2} 1 k} = 0\}$$

$$= \{(j_{1,\nu},\dots,j_{n-2,\nu}) \mid \nu = 1,\dots,|J|\}$$

where the indices $\nu = 1,\dots,|J|$ have been chosen according to a lexicographical ordering of the tuples $(j_1,\dots,j_{n-2})$ for which

$$d_{j_1 \dots j_{n-2} 1 k} = 0$$

for every treatment index k. The set J will again be used to introduce special vertices for the corresponding empty entries of Y (i. e. $y_{j_1 \cdots j_{n-2} 1k} = \emptyset$ with $(j_1, \ldots, j_{n-2}) \in J$) in the digraph to be defined. Using Y and J we get

$$m_{j_1 \cdots j_{n-1}} = \begin{cases} \min \{k \mid k \in y_{j_1 \cdots j_{n-1}}\} & \text{if } y_{j_1 \cdots j_{n-1}} \neq \emptyset ; \\ m_n + \nu & \text{if } y_{j_1 \cdots j_{n-1}} = \emptyset , \ j_{n-1} = 1 \text{ and} \\ & (j_1, \ldots, j_{n-2}) = (j_{1,\nu}, \ldots, j_{n-2,\nu}) \in J ; \\ 0 & \text{if } y_{j_1 \cdots j_{n-1}} = \emptyset \text{ and } j_{n-1} > 1 . \end{cases}$$

Now we are ready to define the digraph $G^n_{2 \ldots n-1}$. The upper index n indicates that the vertices correspond to the levels of the n-th factor (plus special vertices for the elements of J), and the lower indices $2 \ldots n-1$ indicate that the arc labels are chosen according to the indices of the second through the $(n-1)$-th factor. We define

$$G^n_{2 \ldots n-1} = (V,E)$$

by

$$V = \{1, \ldots, m_n + |J|\} ,$$

$$E = \sum_{\nu=1}^{n-1} E_\nu ,$$

where

$$E_1 = \{(k,k^*)_0 \mid \exists \ j_1, \ldots, j_{n-1} \text{ with } k^* = m_{j_1 \cdots j_{n-1}} < k \in y_{j_1 \cdots j_{n-1}}\} ,$$

and for $\nu = 2, \ldots, n-1$

$$E_\nu = \{(k,k^*)_j \mid \exists\ j_1,\dots,j_\nu \text{ with } k = m_{j_1\dots j_\nu 1\dots 1} > 0\ ,$$
$$k^* = m_{j_1\dots j_{\nu-1}1\dots 1} \text{ and } j = \sum_{i=2}^{\nu-1} m_i + j_\nu\ ,\ j_\nu \geq 2\}\ .$$

As usual a vector label $g(\tau) \in \mathbb{Z}^t$ is assigned to each chain τ in the digraph $(t = \sum_{i=2}^{n-1} m_i)$:

$$g(\tau) = \sum_{e_j \in E(\tau^+)} u^j - \sum_{e_j \in E(\tau^-)} u^j\ ,$$

u^j is the j-th unit vector of $\mathbb{R}^t$, $u^0 = 0_t$.

3.5.1 Example

Consider a six-factor design given by a $3 \times 2 \times 2 \times 2 \times 2 \times 12$ matrix $D = ((d_{j_1\dots j_6}))$ with $d_{j_1\dots j_6} \in \{0,1\}$, such that $d_{j_1\dots j_6} = 1$ holds if and only if $(j_1,\dots,j_6)$ is among the following 6-tuples:

(1,1,1,1,1,3) , (1,1,1,1,1,5) , (1,1,1,1,2,7) ,
(1,1,1,2,1,12) , (1,1,2,1,1,2) , (1,1,2,2,1,11) ,
(1,2,1,1,1,4) , (1,2,1,2,1,5) , (1,2,2,1,1,12) ,
(1,2,2,2,1,7) , (2,1,1,1,1,9) , (2,1,2,1,1,8) ,
(2,1,2,2,1,7) , (2,2,1,1,1,10) , (2,2,1,2,1,1) ,
(2,2,2,1,1,2) , (2,2,2,2,1,1) , (3,1,1,1,1,6) ,
(3,1,1,2,1,7) , (3,2,1,1,1,5) .

In the table below this list has been supplemented in lexicographical order by the 6-tuples $(j_{1,\nu},j_{2,\nu},j_{3,\nu},j_{4,\nu},1,m_6+\nu)$, $1 \leq \nu \leq |J|$, indicated by an asterisk. Moreover, the last column of the table gives for each $(j_1,\dots,j_6)$ the arc $(j_6,k)_j$ in G^6_{2345} stemming from this tuple (if any) .

	j_1	j_2	j_3	j_4	j_5	j_6	$(j_6,k)_j$
	1	1	1	1	1	3	-
	1	1	1	1	1	5	$(5,3)_o$
	1	1	1	1	2	7	$(7,5)_8$
	1	1	1	2	1	12	$(12,3)_6$
	1	1	2	1	1	2	$(2,3)_4$
	1	1	2	2	1	11	$(11,2)_6$
	1	2	1	1	1	4	$(4,3)_2$
	1	2	1	2	1	5	$(5,4)_6$
	1	2	2	1	1	12	$(12,4)_4$
	1	2	2	2	1	7	$(7,12)_6$
	2	1	1	1	1	9	-
*	*2*	*1*	*1*	*2*	*1*	*13*	$(13,9)_6$
	2	1	2	1	1	8	$(8,9)_4$
	2	1	2	2	1	7	$(7,8)_6$
	2	2	1	1	1	10	$(10,9)_2$
	2	2	1	2	1	1	$(1,10)_6$
	2	2	2	1	1	2	$(2,10)_4$
	2	2	2	2	1	1	$(1,2)_6$
	3	1	1	1	1	6	-
	3	1	1	2	1	7	$(7,6)_6$
*	*3*	*1*	*2*	*1*	*1*	*14*	$(14.6)_4$
*	*3*	*1*	*2*	*2*	*1*	*15*	$(15,14)_6$
	3	2	1	1	1	5	$(5,6)_2$

*	3	2	1	2	1	16	$(16,5)_6$
*	3	2	2	1	1	17	$(17,5)_4$
*	3	2	2	2	1	18	$(18,17)_6$

The digraph G^6_{2345} given below has the vertices $1,\dots,m_6 + |J|$ and the arcs listed in the above table:

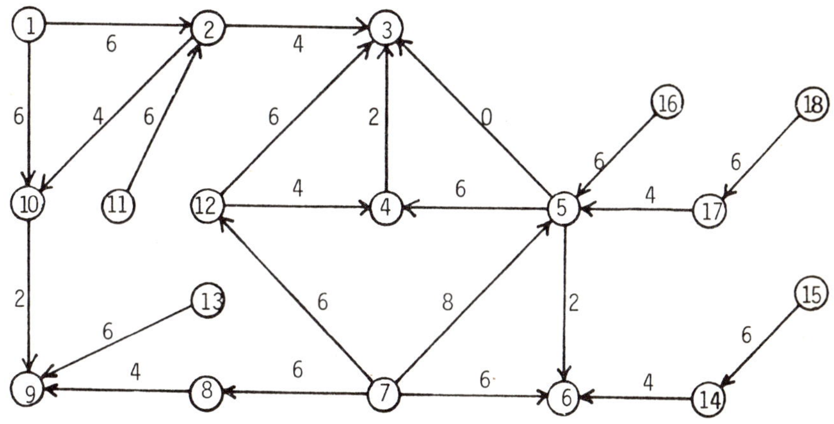

We have $c(G^6_{2345}) = 1$ and $z(G^6_{2345}) = 4$ as the vertex sets $\{1,2,10\}$, $\{3,4,12\}$, $\{3,4,5\}$, $\{5,6,7\}$ determine four independent cycles.

Now consider the $(n-1)$-factor design

$$D_{1\dots n-1} = ((\, d^{1\dots n-1}_{j_1\dots j_{n-1}} \,))$$

induced by an n-factor design D, i. e. let

$$d^{1\dots n-1}_{j_1\dots j_{n-1}} = \sum_{j=1}^{m_n} d_{j_1\dots j_{n-1}j} ,$$

where $j_i = 1,\dots,m_i$ for $i = 1,\dots,n-1$. Using the matrices introduced for (3.1.1) we can apply elementary row operations to the matrices B and B^* of the models corresponding to the designs D and $D_{1\dots n-1}$, resp., yielding

$$P\,A\,B = \begin{bmatrix} I_{m_1} & \tilde{B}_2 & \tilde{B}_3 & \dots & \tilde{B}_n \\ 0_{p-m_1,m_1} & B_2 & B_3 & \dots & B_n \end{bmatrix}$$

$$= \left[P\,A\,B^* , \begin{pmatrix} \tilde{B}_n \\ B_n \end{pmatrix}\right] .$$

In analogy to the definition of the digraph $G^n_{2\dots n-1}$ for the design D we can construct a digraph $G^{n-1}_{2\dots n-2}$ for the design $D_{1\dots n-1}$. It is easy to see that lemma (3.4.3) can be generalized as follows:

$$\operatorname{rank}(B_2,\dots,B_{n-1}) = m_{n-1} - c(G^{n-1}_{2\dots n-2}) + z(G^{n-1}_{2\dots n-2})$$

As a generalization of theorem (3.4.4) we can state:

3.5.2 Theorem

Let D denote an n-factor design and consider the induced (n-1)-factor design $D_{1\dots n-1}$ and the associated digraphs $G^n_{2\dots n-1}$ and $G^{n-1}_{2\dots n-2}$, respectively. Then D is a connected design if and only if $G^n_{2\dots n-1}$ is a connected digraph with

$$z(G^n_{2\dots n-1}) = m_{n-1} - c(G^{n-1}_{2\dots n-2}) + z(G^{n-1}_{2\dots n-2}) .$$

In order to avoid too many technicalities, we omit the proof of the above theorem, which is an immediate extension of the proof of theorem (3.4.4) . It also seems unnecessary to discuss all further connectivity characterizations which result by exchanging the roles of the factors in the construction of the digraph associated to a multi-factor design. Simply exchanging the *blocking* factors does not cause problems, as for symmetry reasons this does not affect the criterion of theorem (3.5.2) . Therefore as a heuristic rule for applying theorem (3.5.2) one should always arrange the blocking factors $F_1,\dots,F_{n-1}$ such that

$$m_1 \geq m_i \geq m_{n-1}, \quad i = 2,\dots,n-2 ,$$

so that only a rather small number of independent cycles has to be found.

We shall now study the relationship between a multi-factor design and its induced subdesigns. Let

$$S \subseteq \{1,\dots,n\} \quad , \quad s = |S| < n \quad ,$$

$$\overline{S} = \{1,\dots,n\} \setminus S ,$$

and consider

$$D_S = ((d^S_{j_1 \dots j_s})) \quad ,$$

$$\text{where } d^S_{j_1 \dots j_s} = \sum_{i \in \overline{S}} \sum_{j_i=1}^{m_i} d_{j_1 \dots j_n} \quad ,$$

i. e. the s-factor design D_S with factors F_i , $i \in S$, induced by an n-factor design D .

3.5.3 Theorem

Let an n-factor design D and $i \in \{1,\dots,n\}$ be given. Consider the (n-1)-factor design D_{S_i} induced by D with $S_i = \{1,\dots,n\} \setminus \{i\}$. Then the following holds: D is a completely connected design if and only if D is F_i-connected and D_{S_i} is a completely connected design.

Proof:

Let B and B^* denote the model matrices for the designs D and $D_{S_n} = D_{1\dots n-1}$ respectively. We have

$$PAB = \left[PAB^* , \begin{pmatrix} \tilde{B}_n \\ B_n \end{pmatrix} \right] ,$$

therefore D is F_n-connected if and only if

$$\text{rank } B = \text{rank } B^* + m_n - 1$$

(cf. lemma (3.1.2)). Theorem (2.3.2) implies that D_{S_n} is a completely connected design if and only if

$$\text{rank } B^* = \sum_{i=1}^{n-1} m_i - (n-1) + 1 \quad .$$

Conversely we know that the inequalities

$$\text{rank } B \leq \text{rank } B^* + m_n - 1 \quad ,$$

$$\text{rank } B^* \leq \sum_{i=1}^{n-1} m_i - n + 2$$

always hold. This proves the theorem for the case $i = n$. For $i \neq n$ the theorem can be verified similarly by exchanging the roles of the factors F_i and F_n . □

If in the experimental design given by D each combination of blocking factor levels occurs with exactly one treatment, then the design can be represented by an an $m_1 \times \dots \times m_{n-1}$ matrix $Y = ((y_{j_1 \dots j_{n-1}}))$ with $y_{j_1 \dots j_n} \in \{1, \dots, m_n\}$,

$$y_{j_1 \dots j_{n-1}} = k \iff d_{j_1 \dots j_{n-1} k} > 0 \quad .$$

The associated digraph becomes somewhat simpler, namely

$$G^n_{2 \dots n-1} = (V,E)$$

$$V = \{1, \dots, m_n\}$$

$$E = \bigcup_{\nu=2}^{n-1} \{ (y_{j_1 \dots j_\nu 1 \dots 1}, y_{j_1 \dots j_{\nu-1} 1 \dots 1})_j \mid j = \sum_{i=2}^{\nu-1} m_i + j_\nu \ ,$$

$$2 \leq j_\nu \leq m_\nu \text{ and } 1 \leq j_i \leq m_i \text{ for } i = 1, \dots, \nu-1 \} \quad .$$

This is a generalization of the row-column designs investigated in section 3.2, and of the four-factor designs treated in corollary (3.4.6). It is easy to check that in this case the digraph $G^{n-1}_{2 \dots n-2}$ of the induced (n-1)-factor design D_{S_n} has the arc set

$$\{ (k,1)_0 \mid 1 \leq k \leq m_3 \} \cup \bigcup_{\nu=2}^{n-2} \{ (1,1)_j \mid j \in \{ \sum_{i=2}^{\nu-1} m_i + j_\nu \mid j_\nu = 2, \dots, m_\nu \} \} \ .$$

It follows that

$$z(G^{n-1}_{2\dots n-2}) = \sum_{i=2}^{n-2} m_i - n + 2 \quad ,$$

i. e. D_{S_n} is completely connected, no matter how the design D of the kind described above has been chosen. This proves the following corollary of theorem (3.5.2).

3.5.4 Corollary

Consider an n-factor design $D = ((d_{j_1 \dots j_n}))$ which is representable by means of an $m_1 \times \dots \times m_n$ matrix $Y = ((y_{j_1 \dots j_{n-1}}))$ with

$$y_{j_1 \dots j_{n-1}} = j \quad \Longleftrightarrow \quad d_{j_1 \dots j_{n-1} j} > 0 \quad .$$

D is a connected design if and only if its associated digraph $G^n_{2\dots n-1}$ is connected with

$$z(G^n_{2\dots n-1}) = \sum_{i=2}^{n-1} m_i - n + 2 \quad .$$

In this case D is even a completely connected design.

We shall now investigate the relations between the connectivity property of a multi-factor design and the connectivity property of certain induced designs.

3.5.5 Theorem

Let D denote an arbitrary n-factor design and consider the induced $|S|$-factor design D_S, where $S \subset \{1,\dots,n\}$ with $n \in S$, $|S| \geq 2$. The following holds: If D is a connected design, then so is D_S.

Proof:

Let Z^i, $i = 1,\dots,n-1$, be a non-singular $p \times p$ matrix such that multiplying the model matrix B of D from the left with Z^i yields a representation of the form

$$Z^i B = \begin{pmatrix} \tilde{B}_1^i & \cdots & \tilde{B}_{i-1}^i & I_{m_i} & \tilde{B}_{i+1}^i & \cdots & \tilde{B}_n^i \\ B_1^i & \cdots & B_{i-1}^i & 0_{p-m_i,m_i} & B_{i+1}^i & \cdots & B_n^i \end{pmatrix}$$

(the corresponding row operations to be applied to B can easily be constructed, cf. also (3.4.7) through (3.4.9)). Furthermore, for a proper subset T of $\{1,\dots,n\}$ with $t = |T| \geq 1$ define a $(p-m_i) \times (\sum_{j\in T} m_j)$ matrix

$$W_T^i = (W_{\nu_1}^i, \dots, W_{\nu_t}^i)$$

where $T = \{\nu_1,\dots,\nu_t\}$ and

$$W_\nu^i = \begin{cases} B_\nu^i & \text{if } \nu \neq i \,, \\ 0_{p-m_i,m_i} & \text{if } \nu = i \,, \end{cases}$$

i. e. W_T^i consists of those submatrices of $(B_1^i,\dots,B_{i-1}^i,0_{p-m_i,m_i},B_{i+1}^i,\dots,B_n^i)$ which correspond to the indices $\nu \in T$. Defining

$$T_o = \{1, \dots, n-1\}$$

the connectivity criterion of lemma (3.1.2) can be rewritten as

$$\text{rank } (W^1_{T_o}, B^1_n) = m_n - 1 + \text{rank } W^1_{T_o} \quad .$$

Analogously one can prove that for any $i \in T_o$ the design D is connected if and only if the condition

$$(3.5.6) \qquad \text{rank } (W^i_{T_o}, B^i_n) = m_n - 1 + \text{rank } W^i_{T_o}$$

is satisfied. Now take any fixed $i^* \in S$, $i^* \neq n$, and let $S_o = S \setminus \{n\}$. Clearly a necessary and sufficient condition for the connectivity of the $|S|$-factor design D_S is

$$(3.5.7) \qquad \text{rank } (W^{i^*}_{S_o}, B^{i^*}_n) = m_n - 1 + \text{rank } W^{i^*}_{S_o} \quad .$$

As $W^{i^*}_{S_o}$ is a submatrix of $W^{i^*}_{T_o}$, condition (3.5.7) is implied by (3.5.6) if one sets $i = i^*$. Therefore the connectivity of D_S follows from the connectivity of D . □

<u>3.5.8 Theorem</u>

All induced ν-factor designs of a completely connected n-factor design are also completely connected $(2 \leq \nu < n)$.

Proof:

Let B be the model matrix of a given n-factor design which is completely connected. Making use of the notation introduced in the proof of the last theorem, we know that for $i \in \{1,\dots,n-1\}$

$$\text{rank } B = \sum_{j=1}^{n} m_j - n + 1 = \text{rank } Z^i B = m_i + \text{rank } W^i_{\{1,\dots,n\}}$$

$$\leq m_i + \sum_{\substack{j=1 \\ j \neq i}}^{n} \text{rank } B^i_j \leq m_i + \sum_{\substack{j=1 \\ j \neq i}}^{n} (m_j - 1)$$

(cf. theorem (2.3.12)). Hence

$$\text{rank } B^i_j = m_j - 1$$

for $i = 1,\dots,n-1$ and $j = 1,\dots,n$ with $i \neq j$. Now consider a subset S of $\{1,\dots,n\}$ with $2 \leq |S| < n$, and let $i^* = \min \{i \mid i \in S\}$. We have

$$\text{rank } W^{i^*}_S = \sum_{i \in S \setminus \{i^*\}} (m_i - 1) = \sum_{i \in S \setminus \{i^*\}} m_i - |S| + 1 \quad ,$$

therefore the design D_S induced by D is also completely connected. □

The converse of theorem (3.5.5) does not hold: this can be seen by the following example.

3.5.9 Example

Consider

$$Y = \begin{bmatrix} 1 & 2 & 3 \\ - & 1 & - \\ - & - & 1 \end{bmatrix}$$

and the three-factor design D determined by

$$d_{ijk} = \begin{cases} 1 & \text{if } y_{jk} = k \ , \\ 0 & \text{otherwise} \ , \quad i,j,k \in \{1,2,3\} \ . \end{cases}$$

Clearly the bipartite graphs G_{12} , G_{23} and G_{13} are connected, hence also the induced two-factor designs D_{12} , D_{23} and D_{13} are (completely) connected (cf. theorem (3.3.19)). But the digraph G (= G_2^3) of theorem (3.3.5)

G_2^3 :

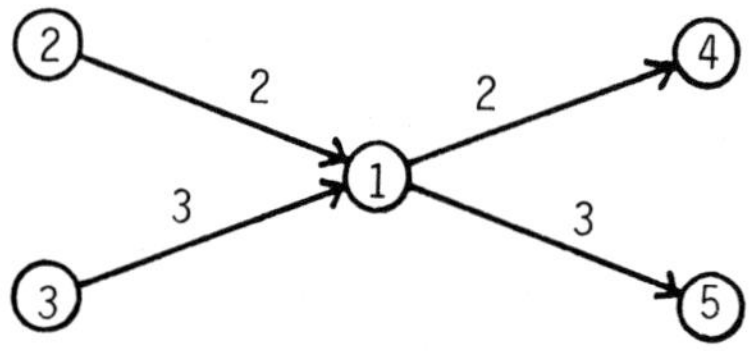

does not contain any cycles, therefore D is not even (F_3-) connected. The digraphs G_2^3 and G_2^1 turn out to be identical for the design in question. We have

G_3^2 :

i. e. D is also neither F_1- nor F_2-connected (cf. (3.3.22)) .

This example contradicts the sufficiency part of a theorem of Ecclestone [1972] stating that the connectivity of all induced two-factor designs is necessary and sufficient for the connectivity of a multi-factor design (the incomplete proof verifies the necessity twice, once directly and once indirectly).

Finally, we shall extend the results of theorem (3.3.27) to the case of n-factor designs. For this purpose some of the notation introduced in section 3.3 has to be modified accordingly. For $i \in \{1,\dots,n\}$ and $\lambda \in \mathbb{R}$ let

$$L = \{f \mid f^T = ((f^1)^T,\dots,(f^n)^T) \text{ , } f^\nu \in \mathbb{R}^{m_\nu} \text{ for } \nu = 1,\dots,n \text{ ,}$$

$$\sum_{j=1}^{m_1} f_j^1 = \dots = \sum_{j=1}^{m_n} f_j^n \} \text{ ,}$$

$$L_i = \{f \mid f \in L, f^\nu = 0_{m_\nu} \text{ for all } \nu \in \{1,\dots,n\} \setminus \{i\} \} \text{ ,}$$

$$\tilde{L}_i = \{f \mid f \in L \text{ , } f^i = 0_{m_i}\} \text{ ,}$$

$$L^\lambda = \{f \mid f \in L \text{ , } \sum_{j=1}^{m_\nu} f_j^\nu = \lambda \text{ for } \nu = 1,\dots,n\} \text{ ,}$$

$$\hat{L} = \{f \mid f \in L \text{ , } f^\nu = u_{m_\nu}^{j_\nu} \text{ , } 1 \le j_\nu \le m_\nu \text{ for } \nu = 1,\dots,n\} \text{ ,}$$

where $u_{m_\nu}^{j_\nu}$ denotes the j_ν-th unit vector of $\mathbb{R}^{m_\nu}$. For $X \subseteq \mathbb{R}^q$, $q = \sum_{\nu=1}^{n} m_\nu$, let

$$M(X)$$

denote the set of all n-factor designs given by a nonnegative integer $m_1 \times \dots \times m_n$ matrix D, for which $\varphi(\pi) = f^T \pi$ is an estimable function for any $f \in X$.

<u>3.5.10 Theorem</u>

Let D be an n-factor design given by a nonnegative integer $m_1 \times \dots \times m_n$ matrix. The following conditions are equivalent for any $\lambda \in \mathbb{R}$, $i \in \{1,\dots,n\}$:

(i) $D \in M(L)$

(ii) $D \in M(L^\lambda)$

(iii) $D \in M(\tilde{L}_i)$

(iv) $D \in M(L_i)$ and the $(n-1)$-factor design $D_{\{1,\dots,n\}\setminus\{i\}}$ induced by D is completely connected.

(v) $D \in M(\hat{L})$

(vi) $D \in M(\bigcup_{\nu=1}^{n} L_\nu)$

<u>Proof:</u>

Consider additionally the conditions

(vii) $D \in M(L^1)$

(viii) $D \in M(L^o)$

(ix) $D \in M(L^1 \cup L^o)$

and the set inclusions

$$M(L) \subseteq M(L^{\lambda}) \qquad (\lambda \in \mathbb{R})$$

$$M(L) \subseteq M(\tilde{L}_i) \qquad (1 \leq i \leq n)$$

$$M(L) \subseteq M(L^1 \cup L^o)$$

$$M(L^o) \subseteq M(\tilde{L}_i) \qquad (1 \leq i \leq n)$$

$$M(L^1 \cup L^o) \subseteq M(L^1)$$

$$M(L^1 \cup L^o) \subseteq M(L^o)$$

$$M(L^1) \subseteq M(\hat{L}) \quad .$$

In the diagram

(3.5.11)

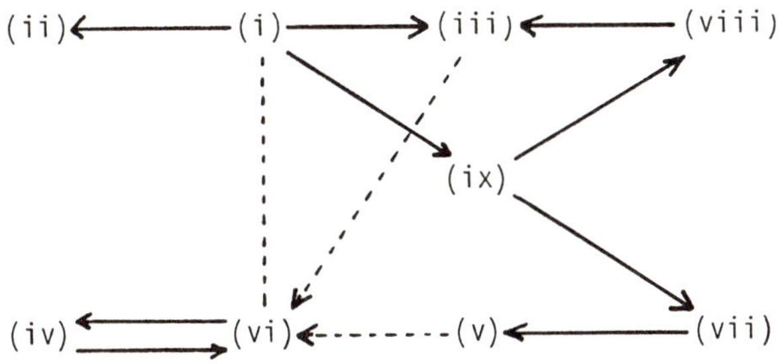

the arrows indicate implications: The equivalence of (iv) and (vi) follows from theorem (3.5.3) , the remaining solid arrows result from the above inclusions; the implications noted by dotted arrows have yet to be verified .

(iii) $\Longrightarrow$ (vi) :

If D is not completely connected, then there is a parameter vector $\pi = ((\pi^1)^T, \dots, (\pi^n)^T)^T$ in the nullspace of the corresponding model matrix

B with

$$\pi^{i^*}_{j'} = \pi^{i^*}_{j''}$$

for some $i^* \in \{1,\dots,n\}$ and $j',j'' \in \{1,\dots,m_{i^*}\}$. We can assume that i^* is different from the index i of condition (iii) : otherwise one can find another index $i' \in \{1,\dots,n\}$, $i' \neq i$, such that for suitable $j_1, j_2 \in \{1,\dots,m_{i'}\}$ $\pi^{i'}_{j_1} \neq \pi^{i'}_{j_2}$ (as each row of $B\pi = 0$ is of the form $\sum_{i=1}^{n} \pi^{i}_{j_i} = 0$ for suitably chosen $j_1,\dots,j_n$ and both $\pi^{i^*}_{j'}$ and $\pi^{i^*}_{j''}$ occur at least once in such an expression, cf. remark (2.1.4)). Let $f \in L$ be defined by

$$f^{\nu}_{j} = \begin{cases} 1 & \text{if } \nu = i^* \text{ and } j = j' , \\ -1 & \text{if } \nu = i^* \text{ and } j = j'' , \\ 0 & \text{otherwise} . \end{cases}$$

Now $i^* \neq i$ guarantees $f \in \tilde{L}_i$, but we have $f^T\pi \neq 0$ for the above $\pi \in$ kernel B . This shows that $D \notin M(\tilde{L}_i)$ if D is not completely connected (i. e. by definition if $D \notin \bigcup_{\nu=1}^{n} L_\nu$) .

(vi) $\Longrightarrow$ (i) :

D is completely connected if and only if for each $\pi = ((\pi^1)^T,\dots,(\pi^n)^T)^T \in$ kernel B

$$\pi^i = \lambda_i \cdot 1_{m_i} \quad , \quad i = 1,\dots,m_i \quad ,$$

for suitable $\lambda_i \in \mathbb{R}$ with $\lambda_n = - \sum_{i=1}^{n-1} \lambda_i$. For each $f \in L$ there is a

$\mu \in \mathbb{R}$ such that $f \in L^\mu$. Consider $\pi \in$ kernel B , $f \in L^\mu$. Then

$$f^T\pi = \sum_{i=1}^{n} (f^i)^T \pi^i$$

$$= \sum_{i=1}^{n-1} \mu\lambda_i + \mu(-\sum_{i=1}^{n-1} \lambda_i) = 0 \quad .$$

(v) $\Longrightarrow$ (vi) :

If condition (vi) is not fulfilled, there is a $\pi \in$ kernel B with

$$\pi^{i^*}_{j'} \neq \pi^{i^*}_{j''}$$

for some $i^* \in \{1,\dots,n\}$, $j',j'' \in \{1,\dots,m_{i^*}\}$. Consider $f,\bar{f} \in \hat{L}$ with

$$f^\nu = \bar{f}^\nu \in \mathbb{R}^{m_\nu} \quad \text{for} \quad \nu \in \{1,\dots,n\} \setminus \{i^*\} \quad ,$$

$$f^{i^*}_j = \begin{cases} 1 & \text{if } j = j' , \\ 0 & \text{otherwise} , \end{cases}$$

$$\bar{f}^{i^*}_j = \begin{cases} 1 & \text{if } j = j'' , \\ 0 & \text{otherwise} . \end{cases}$$

Then $f^T\pi \neq \bar{f}^T\pi$ implies $D \notin M(\hat{L})$.

This proves the equivalence of conditions (i) and (iii) through (ix) (cf. (3.5.11)) . For the case $\lambda = 0$ we have (ii) $\Longleftrightarrow$ (viii) . For $\lambda \neq 0$

$$f \in L^\lambda \quad \Longleftrightarrow \quad \lambda^{-1} f \in L^1$$

yields the equivalence of conditions (ii) and (vii) .

□

The comments following theorem (3.3.27) carry over to the general case, and for general multi-factor designs the property of complete connectivity can also be characterized by means of several equivalent estimability criteria. Theorem (3.5.10) together with lemma (2.1.10) shows that completely connected designs are in some sense "optimal" so far as estimability is connected (cf. Butz [1982]) .

4. INVARIANCE PROPERTIES, REDUCTION METHODS, ALGORITHMS

This last chapter shows how the digraphs introduced for the characterization of connectivity can be handled conveniently. In the corresponding theorems of sections 3.2 through 3.5 certain properties of a given design D (F_i-connectivity or complete connectivity) are characterized by means of certain properties of a digraph which is uniquely determined by D. Each cycle of such a digraph G has a vector label which is determined in a straightforward way by the integer labels of its arcs. The only relevant characteristics of G are the numbers z(G) (maximum number of independent cycles in G, i. e. maximum number of cycles with linearly independent vector labels) and possibly c(G) (number of connected components of G).

We shall first of all investigate ways of modifying (simplifying) a given digraph G subject to the condition that the property of G characterizing the connectivity of the underlying design D remains unchanged. Clearly this invariance condition implies the invariance of the numbers z(G) and c(G). As we want to be able to check connectivity criteria easily (i. e.: determine z(G) and c(G)), the following question must be answered: How can a given digraph G be modified, in particular suitably simplified, without changing z(G) and c(G) ?

By definition z(G) depends on the labels of the cycles in G which are uniquely determined for the digraphs introduced in the preceding chapter. For the more general considerations of this chapter we must emphasize that for each of these digraphs all arcs have labels, which in turn define vector labels for every occurring chain. Here we consider digraphs with integer labeled arcs $((k,k^*)_j, j \in \{0,1,\dots,t\})$ as well as digraphs whose arcs have vector labels $((k,k^*)_r, r \in \mathbb{R}^t)$. Let $\tau_e = [k, (k,k^*)_j, k^*]$ resp. $\tau_e = [k, (k,k^*)_r, k^*]$ denote the chain which consists only of the arc $e = (k,k^*)_j$ resp. $e = (k,k^*)_r$. Then the vector label $g(\tau) \in \mathbb{R}^t$ of an arbitrary chain in $G = (V,E)$ is defined by

$$g(\tau) = \sum_{e \in E(\tau^+)} g(\tau_e) - \sum_{e \in E(\tau^-)} g(\tau_e) ,$$

where $g(\tau_e) = u_t^j$ resp. $g(\tau_e) = r$.

4.1.1 Definition

Let Γ_t denote the set of all arc labeled digraphs with t-dimensional chain labels. A mapping $b: \Gamma_t \longrightarrow \Gamma_t$ is called an *admissible modification* if $z(G)$ and $c(G)$ remain invariant under b (i. e. $z(G) = z(b(G))$ and $c(G) = c(b(G))$ where $G \in \Gamma_t$) . Then the digraph $b(G)$ is *an admissible modification of* $G \in \Gamma_t$ and conversely.

Clearly the statement "G^* is an admissible modification of G " defines an equivalence relation on Γ_t, which will also be denoted by

$$G^* \sim G \quad .$$

For the graphs introduced in chapter 3 we have

$$G_2^3 \sim G_{B_2,B_3} \quad , \quad G_3^2 \sim G_{B_3,B_2} \quad ,$$

$$G_1^3 \sim G_{B_1',B_3'} \quad , \quad G_3^1 \sim G_{B_3',B_1'} \quad ,$$

$$G_1^2 \sim G_{B_1'',B_2''} \quad , \quad G_2^1 \sim G_{B_2'',B_1''} \quad ,$$

cf. section 3.3 , furthermore

$$G_{jk}^i \sim G_{(B_j,B_k),B_i} \qquad \text{for } \{i,j,k\} = \{2,3,4\} \ ,$$

$$G_{jk}^i \sim G_{(B_j^\nu,B_k^\nu),B_i^\nu} \qquad \text{for } \{i,j,k,\nu\} = \{1,2,3,4\} \text{ with } \nu \neq 1 \ ,$$

cf. section 3.4 , and finally

$$G_{2\dots n-1}^n \sim G_{(B_2,\dots,B_{n-1}),B_n} \ ,$$

cf. section 3.5 . For the reduction methods to be described we need the notion of the *contraction* of a set of vertices in a digraph: *contracting* a subset $U \subseteq V$ in $G = (V,E)$ means essentially the identification of all vertices in U with an arbitrary $k_U \in U$ and accordingly "diverting" all those arcs to k_U which are incident to some vertex in U . Thus arcs whose head and tail vertex are both in the set U become loops ; if multiple arcs with the same label occur, only one copy will be kept. Hence the result of the contraction of $U \subseteq V$ in $G = (V,E)$ is the digraph

$$G_U = (V_U, E_U)$$

where

$$V_U = V \setminus (U \setminus \{k_U\}) \qquad (k_U \in U \ \text{arbitrary}) \ ,$$

$$\begin{aligned} E_U = &\{(k,k^*)_j \mid (k,k^*)_j \in E \ , \ k,k^* \in V \setminus U\} \\ &\cup \{(k,k^*)_j \mid \exists \ (k',k^*)_j \in E \ , \ k = k_U \in U \ , \ k' \in U \ , \ k^* \in V \setminus U\} \\ &\cup \{(k,k^*)_j \mid \exists \ (k,k')_j \in E \ , \ k^* = k_U \in U \ , \ k' \in U \ , \ k \in V \setminus U\} \\ &\cup \{(k,k)_j \mid \exists \ (k',k'')_j \in E \ , \ k = k_U \in U \ , \ k',k'' \in U\} \ . \end{aligned}$$

As an example, consider the digraph

G: 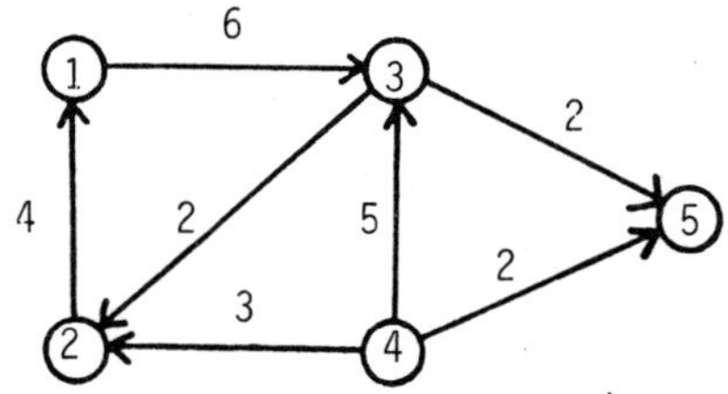 ,

and contract $U = \{3,4\}$, taking $k_U = 3$. The result is the following digraph:

G_U : 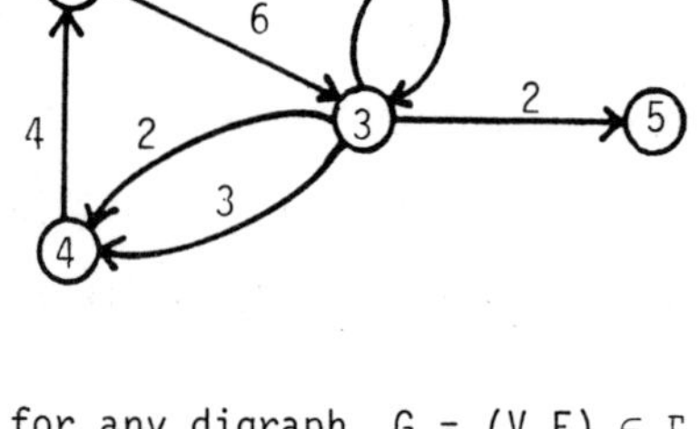 .

It is easy to see that for any digraph $G = (V,E) \in \Gamma_t$ and $U \subseteq V$ we have $G_U \in \Gamma_t$ and

$$c(G) \geqq c(G_U) \quad ,$$

$$z(G) \leqq z(G_U) \quad .$$

4.1.2 Theorem

Consider a digraph $G = (V,E)$ in Γ_t . For chains τ in G let $g(\tau) \in \mathbb{R}^t$ denote the associated vector label. The application of each of the two following *reduction rules* (R1) , (R2) yields an admissible modification of G .

(R1) If there is a chain τ in G with head vertex k , tail vertex k^* and $g(\tau) = 0_t$, contract $U = \{k,k^*\}$.

(R2) If in G there exist a chain τ with head vertex k, tail vertex k^* , and a cycle σ such that $g(\sigma) = \lambda \cdot g(\tau)$ for some $\lambda \neq 0$, contract $U = \{k,k^*\}$.

Proof:

Consider a chain τ with head vertex k , tail vertex k^* , and $g(\tau) = 0_t$. For each cycle σ in $G_{\{k,k^*\}}$ there is a cycle $\tilde{\sigma}$ in G which has the same vector label as σ : if σ passes through k_U $(U = \{k,k^*\})$ then such a $\tilde{\sigma}$ can be obtained by "inserting" the chain τ into σ at k_U . If on the other hand σ does not contain the vertex k_U , σ is a common cycle of G and $G_{\{k,k^*\}}$. Hence we have $z(G_{\{k,k^*\}}) \leq z(G)$ provided that $G_{\{k,k^*\}}$ has been obtained according to (R1) .

Assume that (R2) has been applied and suppose $g(\sigma) \neq 0_t$. Let $\tilde{\sigma}$ denote the cycle in $G_{\{k,k^*\}}$ which is induced by σ (the cycle $\tilde{\sigma}$ *induced* by σ

differs from σ only in the notation of the vertices: k and k^* are always replaced by k_U) . Clearly $g(\sigma) = g(\tilde{\sigma})$. Now choose a maximal set of independent cycles in $G_{\{k,k^*\}}$ which contains $\tilde{\sigma}$ (existence guaranteed by the basis exchange theorem of linear algebra) . Let $\tilde{\sigma},\sigma_2,\dots,\sigma_s$ denote the elements of this set. For each cycle σ_i , $i = 2,\dots,s$, there is a cycle σ_i^* in G such that

$$g(\sigma_i^*) = g(\sigma_i) + \alpha_i\, g(\sigma) ,$$

where

$$\alpha_i = \begin{cases} \lambda^{-1} & \text{if } \sigma_i \text{ contains the vertex } k_U , \\ 0 & \text{otherwise ,} \end{cases}$$

$i = 2,\dots,s$. The g-independence of $\tilde{\sigma},\sigma_2,\dots,\sigma_s$ implies g-independence of $\sigma,\sigma_2^*,\dots,\sigma_s^*$, hence $z(G_{\{k,k^*\}}) \leq z(G)$. In case $g(\sigma) = 0$ one can apply (R1).

Observing that we always have $z(G_U) \geqq z(G)$ and $c(G_{\{k,k^*\}}) = c(G)$ completes the proof.

□

The following list contains interesting special cases of the reduction rules stated in theorem (4.1.2) . Again let $G = (V,E) \in \Gamma_t$.

(R.1.1) 

$\in E \Longrightarrow G \sim G_{\{k,k^*\}}$

(R.1.2) 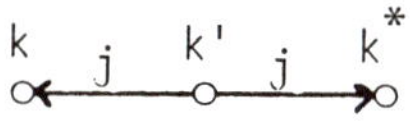

in G $\Longrightarrow G \sim G_{\{k,k^*\}}$

(R.1.3) 

in G such that $g(\tau_1) = g(\tau_2)$

$\Longrightarrow G \sim G_{\{k,k^*\}}$

(R.2.1) 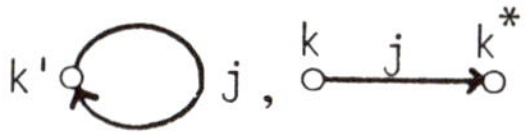

in G $\Longrightarrow G \sim G_{\{k,k^*\}}$

(R.2.2) 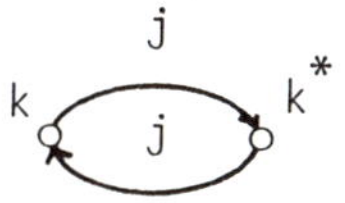

in G $\Longrightarrow G \sim G_{\{k,k^*\}}$

(R.2.3) 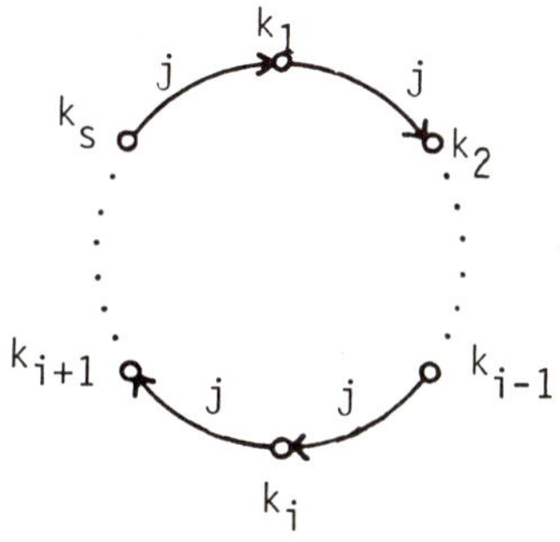

in G $\Longrightarrow G \sim G_{\{k_1,\dots,k_s\}}$

(R.2.4) 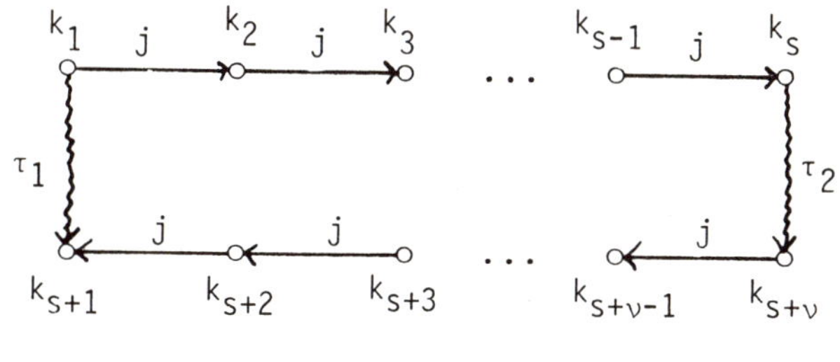

in G

where $g(\tau_1) = g(\tau_2) \Longrightarrow G \sim G_{\{k_1,\dots,k_{s+\nu}\}}$

Frequently one first has to add some arcs in order to be able to apply (R1) or (R2). Adding arcs can be an admissible modification as the following lemma shows (the proof is obvious) .

4.1.2 Lemma

If 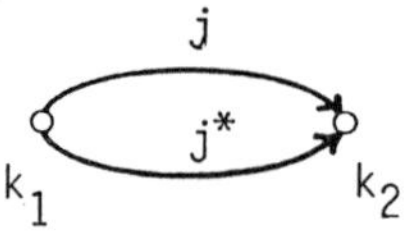

and 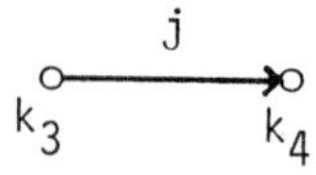

or 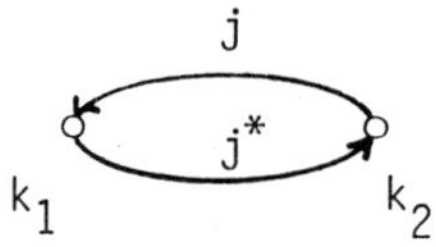

and 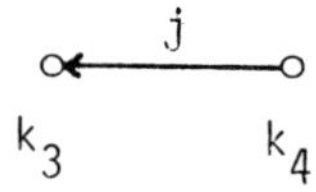

are subgraphs of $G = (V,E) \in \Gamma_t$, then $G^* = (V, E \cup \{(k_3,k_4)_{j*}\})$ is an admissible modification of G .

Several examples taken from the third chapter will now be reviewed to demonstrate the usefulness of our reduction rules. For each contraction to be discussed we shall always take $k_U = \min \{k \mid k \in U\}$. We start with the digraph G_{rc} of example (3.2.1) .

G_{rc} : 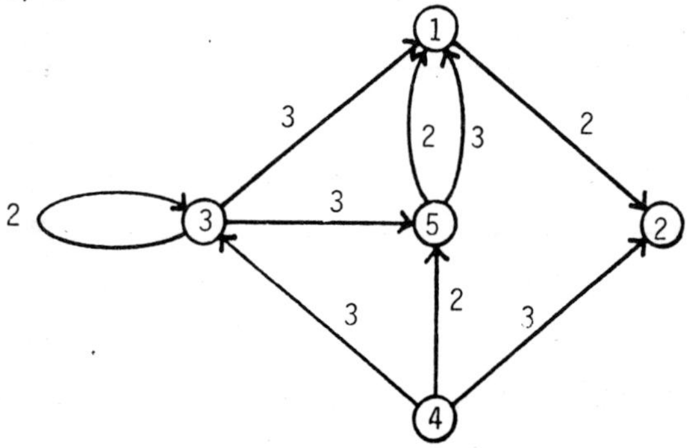

Applying (R.1.2) where $k = 1$, $k' = 3$, $k^* = 5$ yields the admissible modification

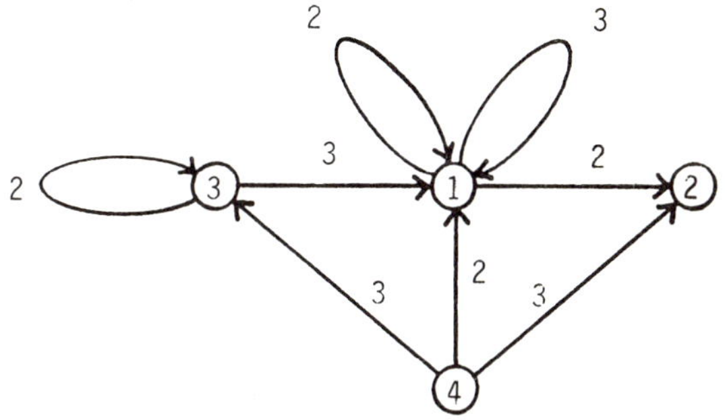

.

By multiple application of (R.2.1) this becomes

,

i. e. we have $z(G_{rc}) = 2$. For the digraph

G :

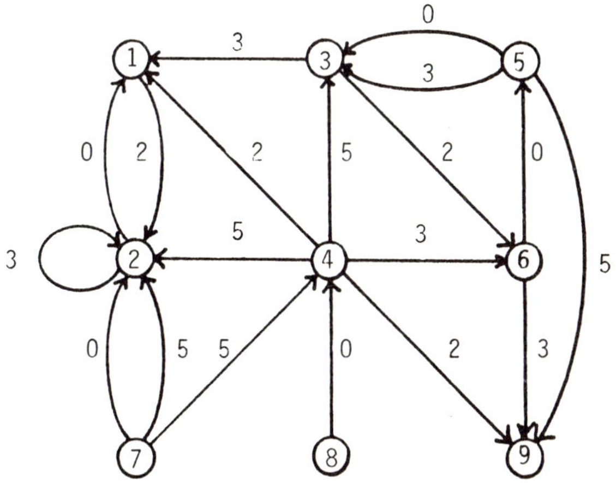

of example (3.4.1) we obtain

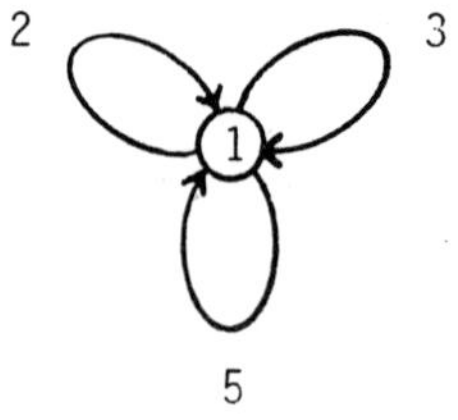

by means of (R.1.1) and (R.2.1) , i. e. $z(G) = 3$.

The digraphs G_2^3 and G_1^3 in example (3.3.18) both reduce to

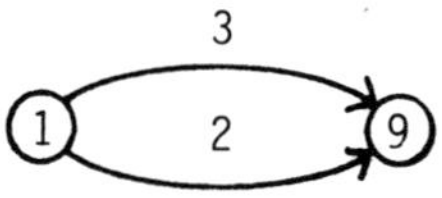

,

and G_{rc} in example (3.2.3) has the admissible modification

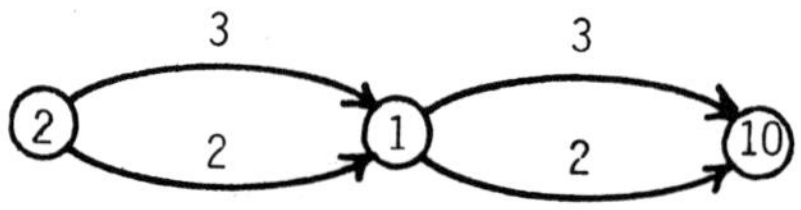

hence $z(G_2^3) = z(G_1^3) = z(G_{rc}) = 1$.

The reduction of G_{2345}^6 taken from example (3.5.1) will now be shown in full detail . Applying (R.1.2) (and the analogous rule obtained by reversing the orientations of the arcs) we get

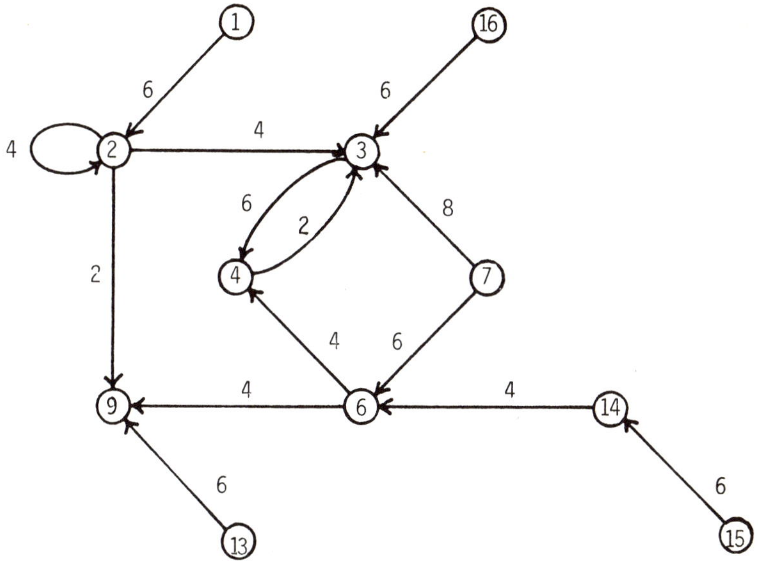

(R.2.1) and again (R.1.2) yield

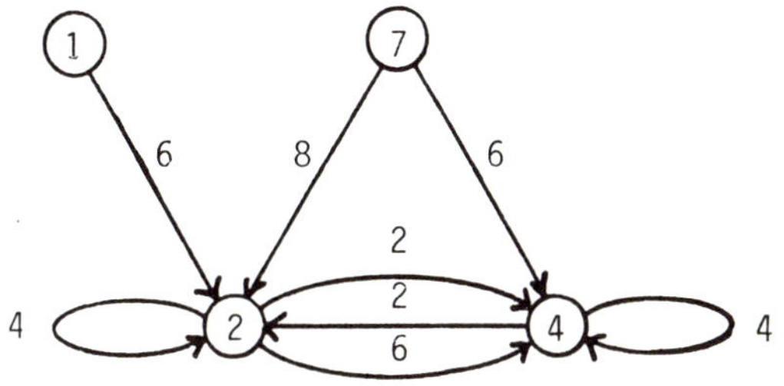

Now by means of (R.2.2) and (R.2.1) one can see that even

is an admissible modification of G^{6}_{2345}, hence $z(G^{6}_{2345}) = 4$.

4.1.4 Theorem

Consider a connected digraph $G = (V,E)$ with arc labels from $\{0,j_1,\dots,j_s\}$. Then we have $z(G) = s$ if and only if G can be reduced to the admissible modification

$$G^* = (\{1\} \;, \{(1,1)_{j_i} \mid i = 1,\dots,s\})$$

by means of the reduction rules of theorem (4.1.2) .

Proof:

The sufficiency part is obvious. To prove necessity, consider s independent cycles $\sigma_1,\dots,\sigma_s$ in G . Without loss of generality we may assume $j_i = i+1$ for $i = 1,\dots,s$. Let $M = (g(\sigma_1),\dots,g(\sigma_s))$ denote the integer $(s+1)\times s$ matrix of vector labels of these cycles. Then we have

$$M = \begin{pmatrix} 0^T_{s+1} \\ \\ A \end{pmatrix}$$

where A is a nonsingular $s\times s$ matrix which has a rational inverse A^{-1} .
The digraph G^* has the following form:

G^* :

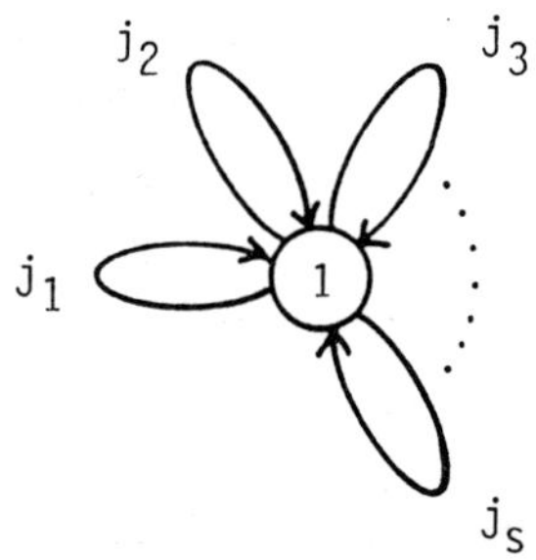

Consider the cycle $\sigma_i^* = [1, (1,1)_{j_i}, 1]$ in G, $i = 1,\dots,s$. Clearly $g(\sigma_i^*) = u_{s+1}^i$. Let

$$\alpha^i = A^{-1} u_s^{i-1} .$$

As

$$M \alpha^i = u_{s+1}^i ,$$

we have

$$g(\sigma_i^*) = \sum_{\nu=1}^{s} \alpha_\nu^i \, g(\sigma_\nu)$$

where α_ν^i denotes the ν-th element of the i-th column of the matrix A^{-1}. Hence for each σ_i^* there is a representation

$$\lambda_i \, g(\sigma_i^*) = \sum_{\nu=1}^{s} \beta_\nu^i \, g(\sigma_\nu)$$

where $\lambda_i \neq 0$ and the β_ν^i are appropriate integers $(i,\nu = 1,\dots,s)$.

Now choose in each cycle σ_i, $i = 1,\dots,s$, in G a vertex $k_i \in V$ (not necessarily distinct). For $i,i^* \in \{1,\dots,s\}$ let τ_{ii^*} denote a chain in G with tail vertex k_i and head vertex k_{i^*}, the existence of such chains is guaranteed by the connectedness of G. From the cycles σ_i and the chains τ_{ii^*} one can construct for each $\nu \in \{1,\dots,s\}$ a cycle $\tilde{\sigma}_\nu$ in G such that $g(\tilde{\sigma}_\nu) = \lambda_\nu u_{s+1}^\nu$: for $\tilde{\sigma}_\nu$ we start at the vertex k_1, pass β_1^ν-times through the cycle σ_1, continue via the chain τ_{12} to the vertex k_2, pass

β_2^ν-times through σ_2 , carry on via τ_{21} and τ_{13} to k_3 , then pass β_3^ν-times through σ_3 , and so on, until after passing β_s^ν-times through σ_s the chain τ_{s1} finally leads back to the starting vertex k_1 of $\tilde{\sigma}_\nu$. The sign of β_i^ν determines in each case whether one passes through σ_i according to the orientation of σ_i (β_i^ν positive) or in the opposite direction (β_i^ν negative) .

Let $\tau_e = [k , (k,k^*)_{j_\nu} , k^*]$ where $e = (k,k^*)_{j_\nu} \in E$. We have $g(\tilde{\sigma}_\nu) = \lambda_\nu \, g(\tau_e)$, therefore by (R2) the set $\{k,k^*\}$ can be contracted. It follows that each pair of vertices in G incident to an arc having one of the positive labels $j_1,\dots,j_s$, can be contracted by means of (R2) . Finally also vertices incident to an arc with label 0 can be contracted (cf. (R1)). The result is the digraph G^* which therefore is an admissible modification of G .

□

The proof of theorem (4.1.4) is not constructive,in the sense that it does not yield a practical method for checking whether $z(G) = s$ holds. Furthermore, in the case of connected but not completely connected designs one has to deal with $z(G) = \ell$ where $\ell < s$. However, the investigation of a great number of examples showed that successive application of the reduction rules usually leads to an admissible modification which easily allows the calculation of the number $z(G)$. Problems may nevertheless occur if, in order to apply (R2) , a cycle has to be found which passes through several elementary cycles, possibly even repeatedly. The following example shows how lemma (4.1.3) can be helpful in such cases.

4.1.5 Example

Consider the following 4×4 set matrix $Y = ((y_{ij}))$ with $y_{ij} = \{k \mid d_{ijk} > 0\}$ and the three-factor design D given by Y:

$$Y = \begin{bmatrix} 3 & 4 & - & - \\ 1 & - & - & 7 \\ 2 & - & 1,4 & 6 \\ 5,6 & 2,7 & - & - \end{bmatrix}$$

The associated digraph G_2^3

G_2^3 :

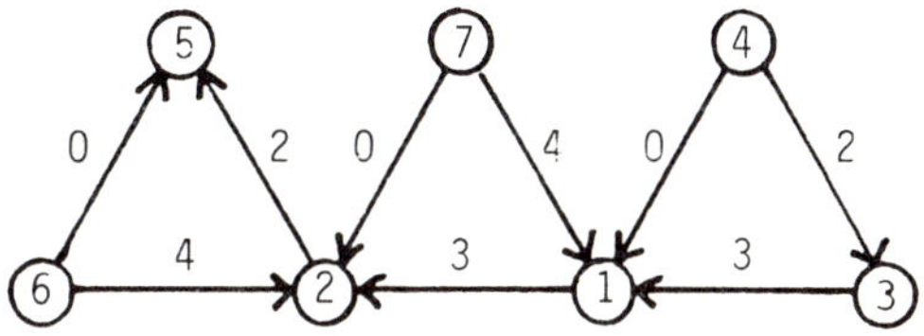

can be reduced by means of (R1) to

G :

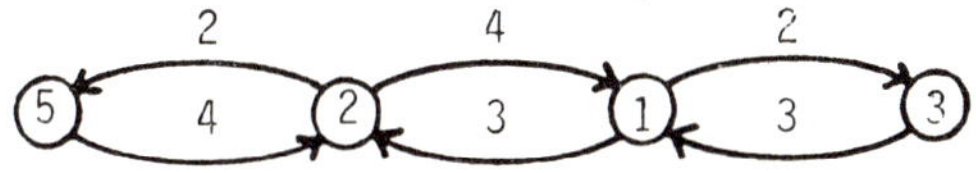

There apparently seems to be no immediate further reduction. However, lemma (4.1.3) yields

G' :

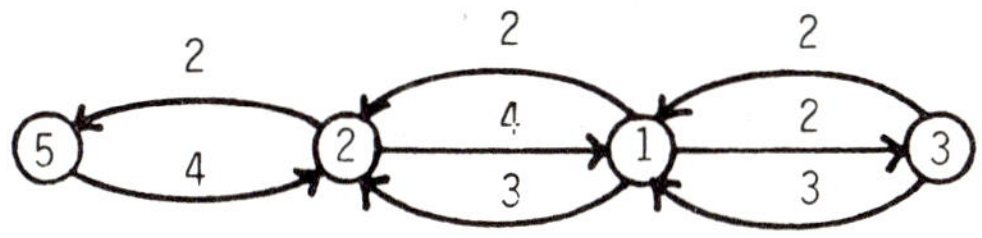

to which (R2) can be applied. The result is

G^* : 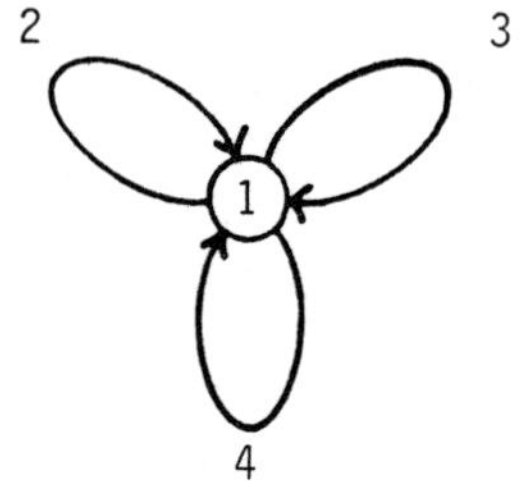

,

i. e. $z(G^*) = z(G') = z(G) = z(G_2^3) = 3$.

The digraphs introduced in the third chapter for the characterization of connectivity have integer arc labels, which makes it very easy to handle the criteria involved. In case a given n-factor design $D = ((d_{j_1 \dots j_n}))$ does not fulfil the condition

$$\sum_{k=1}^{m_n} d_{j_1 \dots j_n} > 0$$

for all $(n-2)$-tuples $j_1, \dots, j_{n-2}$ $(j_i = 1, \dots, m_i \; , \; i = 1, \dots, n-2)$, auxiliary vertices $m_n + 1, \dots, m_n + |J|$ have to be introduced (cf. section 3.5). This may become cumbersome if $|J|$ is a big number. Auxiliary vertices are not necessary if one is willing to deal with higher-dimensional arc labels , cf. the digraph $G_{Q,R}$ in lemma (3.1.5) with $Q = (B_2, \dots, B_{n-1})$ and $R = B_n$ $(B_2, \dots, B_n$ are submatrices of PAB appearing in (3.1.1)).

We now give an algorithm which calculates $z(G^n_{2 \dots n-1})$ using another digraph of the type $G_{Q,R}$. This digraph is also an admissible modification of $G^n_{2 \dots n-1}$.

4.1.6 Algorithm for the calculation of $z(G^n_{2\dots n-1})$

Given: n-factor design $D = ((d_{j_1 \dots j_n}))$ with $d^i_j > 0$ for $i = 1,\dots,n$, $j = 1,\dots,m_i$.

Wanted: $z^D = z(G^n_{2\dots n-1})$.

Let $H = \{(j_1,\dots,j_n) \mid d_{j_1\dots j_n} > 0\}$, $p = |H|$, and $j^1 \lneqq j^2 \lneqq \dots \lneqq j^p$ be a lexicographical ordering of the n-tuples in H.

Furthermore let $w = \sum_{i=2}^{n-1} m_i$ and let $\bar{z}$ denote an upper bound for $z(G^n_{2,\dots,n-1})$, e. g. $\bar{z} = \sum_{i=2}^{n-1} (m_i - 1)$ or $\bar{z} = m_{n-1} - c(G^{n-1}_{2\dots n-2}) + z(G^{n-1}_{2\dots n-2})$.

(1) Initialization:

Set $V := \{1,\dots,m_n\}$, $E := \emptyset$, $G := (V,E)$, $M := I_w$, $z := 0$, $\nu := 1$, $\longrightarrow$ (2).

(2) If $\nu = p$ or $z = \bar{z}$, $\longrightarrow$ (7).

Otherwise $\nu := \nu + 1$, $\longrightarrow$ (3).

(3) If $j_1^\nu = j_1^{\nu-1}$, $\longrightarrow$ (2).

Otherwise

$$r_\nu^T := \left(\left(u^{j_2^{\nu-1}}_{m_2} - u^{j_2^{\nu}}_{m_2} \right)^T, \dots, \left(u^{j_{n-1}^{\nu-1}}_{m_{n-1}} - u^{j_{n-1}^{\nu}}_{m_{n-1}} \right)^T \right),$$

$$e_{r_\nu} := (j_n^\nu, j_n^{\nu-1})_{r_\nu},$$

$$E := E \cup \{e_{r_\nu}\}, \longrightarrow (4).$$

(4) If $G = (V,E)$ does not contain a cycle, $\longrightarrow$ (2) .
Otherwise choose cycle σ in G and calculate

$$g(\sigma) := \sum_{e_r \in E(\sigma^+)} r - \sum_{e_r \in E(\sigma^-)} r ,$$

$\overline{g}(\sigma)$:= vector resulting from $g(\sigma)$ by setting the $(\sum_{i=1}^{j} m_i + 1)$ - th component to 0 for all $j = 0,\dots,n-1$.

$$E := E \setminus \{e_{r_\nu}\} , \longrightarrow (5) .$$

(5) If $\overline{g}(\sigma) = 0_w$, $\longrightarrow$ (2) .
Otherwise $x := M\overline{g}(\sigma)$, $\longrightarrow$ (6) .

(6) If $x = 0_w$, $\longrightarrow$ (2) .
Otherwise $j^* := \min \{j \mid x_j \neq 0\}$.
Define a $w \times w$ matrix $M^x = ((m_{ij}^x))$ by

$$m_{ij}^x = \begin{cases} 1 & \text{if } i = j \neq j^* , \\ 0 & \text{if } i \neq j \neq j^* \text{ or } i = j = j^* , \\ -x_i/x_j & \text{if } i \neq j = j^* , \end{cases}$$

$$M := M^x M , \; z := z + 1 , \longrightarrow (2) .$$

(7) $z^D := z$, stop.

Comments on the algorithm (4.1.6):

The set $H = \{j^1, \dots, j^p\}$ contains all n-tuples $j^\nu = (j_1^\nu, \dots, j_n^\nu)$ with $d_{j_1^\nu \dots j_n^\nu} > 0$, the upper indices being chosen according to a lexicographical ordering of these tuples. The algorithm successively produces arcs with vector labels of a digraph of the type $G_{Q,R}$. The digraph $G_{(B_2,\dots,B_{n-1}),B_n}$ used in the last chapter refers to

$$P\,A\,B = \begin{bmatrix} I_{m_1} & \tilde{B}_2 & \tilde{B}_3 & \dots & \tilde{B}_n \\ 0_{p-m_1,m_1} & B_2 & B_3 & \dots & B_n \end{bmatrix}$$

(B is the model matrix of the design in question) , whereas for the digraph G constructed in the algorithm we have

$$G = G_{(\bar{B}_2,\dots,\bar{B}_{n-1}),\bar{B}_n}$$

where $\bar{B}_2, \dots, \bar{B}_n$ are submatrices of

$$(4.1.7) \qquad P\,\bar{A}\,B = \begin{bmatrix} I_{m_1} & \tilde{B}_2 & \tilde{B}_3 & \dots & \tilde{B}_n \\ 0_{p-m_1,m_1} & \bar{B}_2 & \bar{B}_3 & \dots & \bar{B}_n \end{bmatrix} .$$

The permutation matrices P of (3.1.1) and (4.1.7) are the same. As explained in section 3.1 , multiplying B from the left with P A causes elementary row operations. Instead of A we have in (4.1.7) the $p \times p$ block-diagonal matrix

$$\bar{A} = \begin{bmatrix} \bar{A}_1 & & & & 0 \\ & \bar{A}_2 & & & \\ & & \cdot & & \\ & 0 & & \cdot & \\ & & & & \cdot \\ & & & & \bar{A}_{m_1} \end{bmatrix}$$

on whose diagonal are $d_j^1 \times d_j^1$ matrices

$$\bar{A}_j = \begin{bmatrix} 1 & & & & & 0 \\ 1 & -1 & & & & \\ & 1 & -1 & & & \\ & & \cdot & \cdot & & \\ & 0 & & \cdot & \cdot & \\ & & & & \cdot & \cdot \\ & & & & 1 & -1 \end{bmatrix}$$

$j = 1,\dots,m_1$. It is easy to see that the first m_1 rows of $P\,A\,B$ and of $P\,\bar{A}\,B$ are identical. According to definition (4.1.1) we have

$$G_{(\bar{B}_2,\dots,\bar{B}_{n-1}),\bar{B}_n} \sim G_{(B_2,\dots,B_{n-1}),B_n} \sim G^n_{2\dots n-1}$$

hence in particular $z(G) = z(G^n_{2\dots n-1})$. Each row of $(\bar{B}_2,\dots,\bar{B}_n)$ corresponds to an arc of G and identical rows are associated with the same arc. By definition of $G_{R,Q}$ (cf. section 3.1) a row (r,q) of $[(\bar{B}_2,\dots,\bar{B}_{n-1}),\bar{B}_n]$ with

$$q^T = u_{m_n}^{j_n^{\nu-1}} - u_{m_n}^{j_n^{\nu}}$$

yields the arc $(j_n^\nu, j_n^{\nu-1})_r$ (if $j_n^\nu = j_n^{\nu-1}$ there are further loops with the label r, but they are irrelevant); by construction of $P\bar{A}B$ such a row exists only for $\nu = 2,\dots,p$ with $j_1^\nu = j_1^{\nu-1}$.

The algorithm (4.1.6) successively produces and evaluates the arcs of this digraph $G_{R,Q}$. In step (3) the arc e_{r_ν} is added to G if $j_1^{\nu-1} = j_1^\nu$ for $j^{\nu-1}, j^\nu \in H$. Every time a new arc e_{r_ν} closes a cycle σ in G, this arc is again eliminated after $g(\sigma)$ has been calculated, cf. (4). This implies that in step (2) of the algorithm G is always cycle-free and that at the beginning of step (4) G contains either exactly one elementary cycle (uniquely determined up to the orientation) or none.

$g(\sigma)$ is a linear combination of the rows of $(\bar{B}_2,\dots,\bar{B}_{n-1})$ and can accordingly be subdivided into subvectors $g^i(\sigma) \in \mathbb{Z}^{m_i}$: $g(\sigma) = (g^2(\sigma)^T,\dots,g^{n-1}(\sigma)^T)^T$ with $g^i(\sigma) = (g_1^i(\sigma),\dots,g_{m_i}^i(\sigma))^T$ where $i = 2,\dots,n-1$. Now let for $i = 2,\dots,n-1$, $j = 1,\dots,m_i$

$$\bar{g}_j^i(\sigma) = \begin{cases} 0 & \text{if } j = 1, \\ g_j^i(\sigma) & \text{if } 2 \leq j \leq m_i, \end{cases}$$

and $\bar{g}^i(\sigma) = (\bar{g}_1^i(\sigma),\dots,\bar{g}_{m_i}^i(\sigma))^T$, $\bar{g}(\sigma) = (\bar{g}^2(\sigma)^T,\dots,\bar{g}^{n-1}(\sigma)^T)^T$. By construction we have $1_{m_i}^T \cdot g^i(\sigma) = 0$ for all $i \in \{2,\dots,n-1\}$, hence ℓ cycles are g-independent if and only if they are $\bar{g}$-independent. Therefore one can use $\bar{g}(\sigma)$ in the algorithm after a cycle has been identified in step (4). Steps (5) and (6) contain a simple elimination technique for the calculation of $z^D = z(G_{R,Q})$ by means of the vectors $\bar{g}(\sigma) \in \mathbb{Z}^W$ generated in step (4). The algorithm starts with $M = I_w$, and the first cycle σ_1 with $\bar{g}(\sigma_1) \neq 0_w$ yields $M = M^{\bar{g}(\sigma_1)} \cdot I_w$ and $z = 1$ (counter for z^D, cf. (6)). For the next cycle σ_2 we have

$$x = M\,\bar{g}(\sigma_2) = M^{\bar{g}(\sigma_1)}\,\bar{g}(\sigma_2) \neq 0_w$$

if and only if $\bar{g}(\sigma_1)$, $\bar{g}(\sigma_2)$ are linearly independent. In the case $x \neq 0_w$ we get $M = M^x M^{\bar{g}(\sigma_1)}$ and $z = 2$. Then we have $M\,\bar{g}(\sigma_3) \neq 0_w$ if and only if $\bar{g}(\sigma_1)$, $\bar{g}(\sigma_2)$, $\bar{g}(\sigma_3)$ are linearly independent, and so on. More generally: let $x,y \in \mathbb{R}^w$, $j^* = \min\{j \mid x_j \neq 0\}$, then

$$M^x y = y - y_{j^*}(x_{j^*})^{-1} x$$

by definition of M^x. It is easy to verify that

$$M^{x^k} M^{x^{k-1}} \ldots M^{x^1} x^{k+1} \neq 0_w$$

holds if and only if $\operatorname{rank}(x^1,\ldots,x^{k+1}) = k + 1$.

In order to determine whether the digraph under consideration in step (4) has a cycle, one must successively calculate $g(\tau_{k,k^*})$ for all chains τ_{k,k^*} with tail vertex k and head vertex k^* ($k < k^*$, $k,k^* \in V$). This is done by a simple book-keeping procedure whenever a new arc enters the digraph. If in step (3) an arc e_{r_ν} is added whose head and tail vertex are already linked by some chain in G, then r_ν and the vector label of this chain immediately yield the vector label of the resulting cycle. For the calculation of all possible $g(\tau_{k,k^*}) \in \mathbb{Z}^w$ $(k,k^* \in \{1,\ldots,m_n\}$, $k < k^*)$ at most $m_n(m_n - 1)w/2$ elementary operations are needed. Calculating $M\,\overline{g}(\sigma)$ in step (5) and $M^x M$ in step (6) requires $O(w^2)$ operations each (O denotes Landau's symbol), and step (5) is carried out less than p times and step (6) less than w times during the algorithm. Thus a rough estimate of the time complexity of (4.1.6) is

$$O(m^3 n + pw^2 + w^3) = O(pw^2) \quad ,$$

where $m = \max\{m_i \mid i = 1,\ldots,n\}$, $p \geq w$.

4.1.7 Example

Consider a five-factor design given by a $10 \times 4 \times 2 \times 3 \times 6$ matrix $D = ((d_{j_1 \cdots j_5}))$ with $d_{j_1 \cdots j_5} \in \{0,1\}$. Let $d_{j_1 \cdots j_5} = 1$ if and only if $(j_1,\ldots,j_5)$ appears in the following list:

(1,1,1,1,1) , (1,2,1,1,3) , (1,2,1,2,6) , (1,2,2,1,4) ,
(1,2,2,3,3) , (2,1,1,1,4) , (2,1,2,1,5) , (2,2,2,1,3) ,
(3,1,1,3,1) , (3,4,1,3,2) , (3,4,2,3,1) , (4,1,1,1,1) ,
(4,3,1,1,3) , (5,3,2,2,5) , (5,4,2,2,3) , (6,3,2,2,5) ,
(6,4,2,2,4) , (6,4,2,3,1) , (7,1,1,2,2) , (7,3,2,2,3) ,
(8,1,1,3,3) , (9,2,2,2,5) , (9,3,1,1,2) , (9,4,2,3,3) ,
(9,4,2,3,1) , (10,1,1,1,2) , (10,1,2,2,5) , (10,3,2,2,2) .

In applying the algorithm described under (4.1.6) it turns out that the analysis of the first eleven arcs suffices to prove that D is completely connected. Only these eleven arcs are shown in the figure:

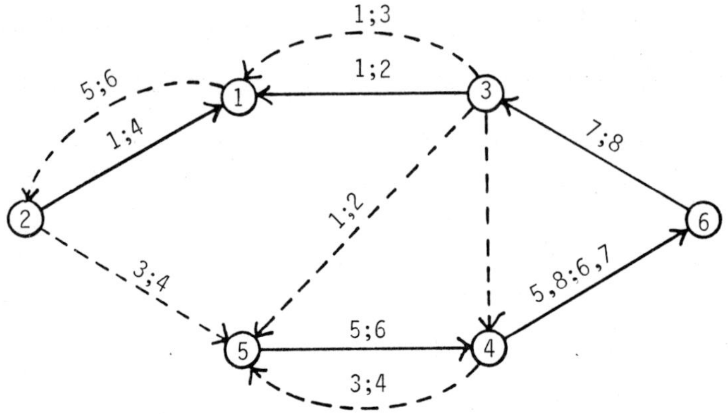

The vector labels r_ν of these arcs are given in short-hand notation: i;i' denotes $u^i - u^{i'}$ and i,j;i',j' denotes $u^i + u^j - u^{i'} - u^{j'}$. Arcs which close a cycle and thus get eliminated after the evaluation of the cycle (step (4)) are drawn as dotted lines. We give a brief indication of these evaluations in chronological order of the occurring cycles.

$(3,4)_{7;9}$: $x = \overline{g}(\sigma) = (0,0,0,0,0,-1,0,0,-1)^T$, $z = 1$

$(3,5)_{1;2}$: $\overline{g}(\sigma) = (0,-1,0,0,0,-2,0,0,0)^T$,
$x = (0,-1,0,0,0,0,0,0,0,2)^T$, $z = 2$.

$(1,2)_{5;6}$: $\overline{g}(\sigma) = (0,0,0,-1,0,-1,0,0,0)^T$,
$x = (0,0,0,-1,0,0,0,0,1)^T$, $z = 3$.

$(3,1)_{1;3}$: $\bar{g}(\sigma) = (0,1,-1,0,0,0,0,0,0)^T$,
$x = (0,1,-1,0,0,0,0,0,0)^T$, $z = 4$.

$(2,5)_{3;4}$: $\bar{g}(\sigma) = (0,-1,1,0,0,-2,0,0,0)^T$,
$x = (0,0,0,0,0,0,0,0,2)^T$, $z = 5$.

$(4,5)_{3;4}$: $\bar{g}(\sigma) = (0,0,1,-1,0,-1,0,0,0)^T$,
$x = (0,0,1,0,0,0,0,0,0)^T$, $z = 6$.

The last 5-tuple to be dealt with is (6,4,2,2,4) . The digraph is connected and $z^D = 6 = \sum_{i=2}^{n-1} (m_i - 1)$, hence D is a completely connected design (i. e. no further 5-tuples need to be considered).

As a rule one will make use of the algorithm (4.1.6) only when one is forced by the dimensions of the design in question to use a computer for checking the connectivity property. For many real world problems the digraphs introduced in the third chapter together with the reduction method of this chapter suffice for an easy analysis.

The connectivity properties of specially structured designs (including very large ones) can often be studied very easily by means of the associated digraphs. We demonstrate this in the following example taken from Weeks/Williams [1964] , which has a straightforward generalization to designs with arbitrarily many factors and factor levels.

4.1.8 Example

Suppose a three-factor design is given by a $5 \times 5 \times 5$ matrix D with

$$d_{ijk} = \begin{cases} 1 & \text{if } k \in y_{ij} , \\ 0 & \text{otherwise} , \end{cases}$$

where

$$y_{ij} = \begin{cases} \{2,4\} & \text{if } i+j \text{ is even} , \\ \{1,3,5\} & \text{if } i+j \text{ is odd} . \end{cases}$$

G_2^3 :

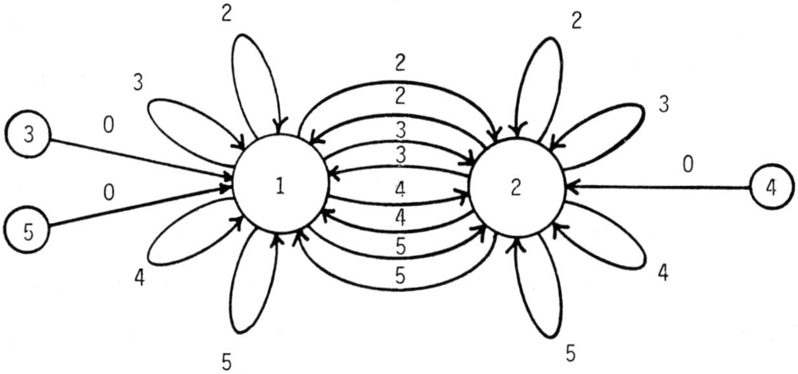

By mere inspection of the digraph G_2^3 it is clear that D is a completely connected design.

In practical applications one is frequently faced with the problem of extending a given design to a connected or even completely connected design by adding further experiments. Or conversely one would like to know which of the experiments are dispensable in the sense that they do not effect the connectivity properties of the design. Furthermore, for theoretical considerations it is often desirable to be able to construct special examples and understand their local connectivity behaviour. For all these and many related problems the combinatorial approach is able to offer useful tools.

LIST OF SYMBOLS

symbol	introduced on page
B	7
$c(G)$	23
$c^{(1)}$	25
$c^{(2)}$	31
$c^{(2)}_{rc}$	32
d^{i}_{j}	16
$d^{ii^*}_{j_i j_{i^*}}$	17
$d_{j_1 \ldots j_n}$	9
diag ( )	16
D	9
D_i	16
D_{ii^*}	17
$D_{1 \ldots n-1}$	139
D_S	141
E	22
$\mathcal{E}$	8
$E(\tau^+)$	24
$E(\tau^-)$	24
ε	7
f	13
f^i	15
f^i_j	15
F_i	7
$G = [V,E]$	22
$G = (V,E)$	23
G_Q	42
$G_{R,Q}$	45
$G_{12} = [V_{12},E_{12}]$	77
$G_{23} = [V_{23},E_{23}]$	96
$G_{13} = [V_{13},E_{13}]$	96
G^1_2	88
G^2_1	87
G^2_3	79
G^3_2	79
G^1_3	84
G^3_1	83
G^i_{jk}	128
$G^n_{2 \ldots n-1}$	135
G_U	157
Γ_t	156
I_m	32
$M(X)$	106
m_i	7
$\mathbb{N}$	44

symbol	introduced on page
$\mathbb{N}_o$	44
1_m	13
$1_{m,m^*}$	32
π	7
π^i	7
$\pi^i_{j_i}$	7
p	7
q	7
$\mathbb{R}^m$	8
$(\ )^T$	7
τ	22
u^k	43
V	22
x	7
X^-	30
Y	11,59,135
z(G)	99
$\mathbb{Z}^m$	53
0_m	15
$0_{m,m^*}$	41

REFERENCES

R.C. Bose [1947] : The design of experiments, *Proceedings of the* 34th *Indian Science Congress* (1947) 1 - 25.

R.C. Bose [1950/51] : *Least Square Aspects of Analysis of Variance*, Mimeo Series No. 9, Institute of Statistics, University of North Carolina 1950/51.

L. Butz [1980] : *Characterizations of Connectivity in Row-Column Designs*, Report No. 80153 - OR, Institut für Ökonometrie und Operations Research, Universität Bonn 1980.

L. Butz [1981] : Connectivity in general designs with two blocking factors, *Methods of Operations Research* 41 (1981) 129 - 132.

L. Butz [1982] : A note on estimability characterizations of completely connected designs for multi-way elimination of heterogeneity, Report No. 81184 - OR, Institut für Ökonometrie und Operations Research, Universität Bonn 1981, to appear in *Zeitschrift für Angewandte Mathematik und Mechanik* 62 (1982).

M.C. Chakrabarti [1962] : *Mathematics of Design and Analysis of Experiments*, Bombay 1962.

C.-S. Cheng [1978] : Optimal designs for the elimination of multi-way heterogeneity, *Annals of Statistics* 6 (1978) 1262 - 1272.

J.A. Ecclestone [1972] : *On the Theory of Connected Designs*, Dissertation, Cornell University 1972.

J. Ecclestone and K. Russell [1975] : Connectedness and orthogonality in multi-factor designs, *Biometrika* 62 (1975) 341 - 345.

W.T. Federer, R.C. Nair and D. Raghavarao [1975] : Some augmented row-column designs, *Biometrics* 31 (1975) 361 - 374.

W. Federerer and M. Zelen [1964] : Applications of the calculus for factorial arrangements II: Two-way elimination of heterogeneity, *Annals of Mathematical Statistics* 35 (1964) 658 - 672.

N. Gaffke [1978] : *Optimale Versuchsplanung für lineare Zwei-Faktor-Modelle*, Dissertation, RWTH Aachen 1978.

W. Y. Gateley [1962] : *Application of the Generalized Inverse Concept to the Theory of Linear Statistical Models*, Dissertation, Oklahoma State University 1962.

F.A. Graybill [1961] : *An Introduction to Linear Statistical Models I*, New York 1961.

B. Jones [1979] : Algorithms to search for optimal row-and-column designs, *Journal of the Royal Statistical Society* B 41 (1979) 210 - 216.

J. Kiefer [1959] : Optimum experimental designs, *Journal of the Royal Statistical Society* B 21 (1959) 272 - 319.

O. Krafft [1978] : *Lineare statistische Modelle und optimale Versuchspläne*, Göttingen 1978.

V.G. Kurotschka and P.S. Dwyer [1974] : Optimal design of three way layouts without interaction, *Mathematische Operationsforschung und Statistik* 5 (1974) 131 - 145.

S.C. Pearce [1975] : Row-and-column designs, *Applied Statistics* 24 (1975) 60 - 74.

D. Raghavarao [1971] : *Constructions and Combinatorial Problems in Design of Experiments*, New York 1971.

D. Raghavarao and W.T. Federer [1975] : On connectednedd in two-way elimination of heterogeneity designs, *Annals of Statistics* 3 (1975) 730 - 735.

K.G. Russell [1976] : The connectedness and optimality of a class of row-column designs, *Communications in Statistics - Theory and Methods* A 5 (1976) 1479 - 1488.

S.R. Searle [1971] : *Linear Models*, New York 1971.

K.R. Shah and C.G. Khatri [1973] : Connectedness in row-column designs, *Communication in Statistics* 2 (1973) 571 - 573.

S.S. Shrikhande [1951] : Designs for two-way elimination of heterogeneity, *Annals of Mathematical Statistics* 22 (1951) 235 - 247.

M. Singh and A. Dey [1978] : Two-way elimination of heterogeneity, *Journal of the Royal Statistical Society* B 40 (1978) 58 - 63.

J.N. Srivastava and D.A. Anderson [1970] : Some basic properties of multidimensional partially balanced designs, *Annals of Mathematical Statistics* 41 (1970) 1438 - 1445.

K.D. Tocher [1952] : The design and analysis of block experiments, *Journal of the Royal Statistical Society* B 14 (1952) 45 - 100.

D.L. Weeks and D.R. Williams [1964] : A note on the determination of connectedness in an N-way cross classification, *Technometrics* 6 (1964) 319 - 324.
Errata: *Technometrics* 7 (1965) 281.

H.P. Wynn [1977] : The combinatirial characterization of certain connected $2 \times J \times K$ three-way layouts, *Communications in Statistics - Theory and Methods* A 6 (1977) 945 - 953.

F. Yates [1936] : Incomplete randomized blocks, *Annals of Eugenics* 7 (1936) 121 - 140.

INDEX

adjacent 22
admissible modification 156
Anderson, D.A. 13,20,187
arc 23
balanced cycle 24
bipartite graph 23
block design 10,11,25,27
blocking effect 10
blocking factor 10
blocking level 10
Bose, R.C. 2,12,20,26,27,185
Butz, L. 54,67,154,185
C-matrix 2,23,31,32,36
chain 22
Chakrabarti, M.C. 2,13,25,32,185
Cheng, C.-S. 3,13,36,185
completely connected design 36
complete graph 22
connected component 23
connected design 20
connected graph 22
contraction 157
contrast 19
cycle 22
design 7
design for two-way elimination of heterogeneity 10,30,59
Dey, A. 13,187
digraph 23
Dwyer, P.S. 13,186
Ecclestone, J.A. 13,20,38,149,185
edge 22
elementary contrast 20
elementary cycle 22
estimable 13
experimental design 7
factor 7
factor levels 7
Federer, W.T. 2,13,31,58,99,102,185,186
F_i-connected 20
F_i-contrast 19
four-factor design 113
Gaffke, N. 3,27,178
Gateley, W.Y. 36,37,186
graph 22
Graybill, F.A. 20,186
head vertex 22
incident 22
independent 45
induced design 96,141

interactions 9
isolated vertex 22
Jones, B. 13,186
Khatri, K.R. 13,58,187
Kiefer, J. 12,186
Krafft, O. 7,12,13,27,186
Kurotschka, V.G. 13,186
labeling function 45
linear estimation 13
linear statistical model 7
loop 23
multi-factor design 12,135
n-factor design 9
Nair, R.C. 13,185
observation 7
observational error 7
partial graph 23
Pearce, S.C. 13,186
permutation matrix 40
Raghavarao, D. 2,12,13,31,99,102
185,186
reduction rules 159,161
row-column design 11,31,51
Russell, K.G. 13,20,185,187
Searle, S.R. 7,187
Shah, K.R. 13,58,187
Shrikhande, S.S. 13,187
Singh, M. 13,187
spanning tree 23
Srivastava, J.N. 13,20
subgraph 22
tail vertex 22
testable linear hypotheses 21
three-factor design 30,59
Tocher, K.D. 12,187
tree 23
vertex 22
Weeks, D.L. 20,36,37,38,179,187
Williams, D.R. 20,36,37,38,179,187
Wynn, H.P. 3,32,33,35,58,187
Yates, F. 2,10,187
Zelen, M. 58,186